Hexaferrite Permanent Magnetic Materials

by
Sami H. Mahmood
Ibrahim Abu-Aljarayesh

The relatively high metallicity of magnetic materials for practical applications imposes limitations for their efficient use due to their unfavorable characteristics. Accordingly, magnetic oxides with ferromagnetic properties emerged as the most widely used magnetic materials for practical applications, owing to their characteristic high resistivity and low eddy current losses, chemical stability, simplicity of production in mass quantities, and other favorable characteristics. An important class of these oxides is the class of hexagonal ferrites developed in the early 1950's, which dominated the world market of permanent magnet applications since the end of the 1980's. Among these ferrites, the magnetoplumbite (M-type) hexaferrite, is produced nowadays in large quantities at very competitive low prices, thus providing the permanent magnet market with probably the most cost-effective magnetic material.

This concise book is intended to provide an overview of the basic concepts of magnetism and magnetic properties pertinent to permanent magnetic materials, importance of these materials in terms of their market share and versatility of practical use, synthesis techniques, and routes adopted for the modification and tuning of their magnetic properties. Emphasis is placed on hexaferrite materials for permanent magnet applications, with M-type ferrites as the focal point. The discussion is kept brief, in an attempt to provide a wide spectrum of knowledge for quick reference to specialized scientists and engineers in this ever increasing industry.

Hexaferrite Permanent Magnetic Materials

By

Sami H. Mahmood[1] and Ibrahim Abu-Aljarayesh[2]

[1]The University of Jordan, Amman, Jordan
[2]Yarmouk University, Irbid, Jordan

Published by **Materials Research Forum LLC**
Millersville, PA 17551, USA

Published as part of the book series
Materials Research Foundations
Volume 4 (2016)
ISSN 2471-8890 (Print)
ISSN 2471-8904 (Online)

Print ISBN 978-1-945291-06-7
ePDF ISBN 978-1-945291-07-4

Distributed worldwide by

Materials Research Forum LLC
105 Springdale Lane
Millersville, PA 17551
USA
http://www.mrforum.com/

Manufactured in the United State of America
10 9 8 7 6 5 4 3 2 1

Table of Contents

Foreword

The utilization of magnetic materials, especially those with ferromagnetic or ferrimagnetic properties, in a wide range of industrial and technological applications, have contributed largely to convenience and prosperity in our modern life. The range of applications of these materials have extended from their limited use in compasses in the ancient era, and in lifting iron scrap pieces in the eighteenth century, to almost uncountable number of devices and machines used in the automated industries, power conversion, traction, electronics and mechatronics, home and office appliances, telecommunications, space technology, and medical applications, to mention a few. Such applications demanded great efforts dedicated to the development of new magnetic materials with tunable properties to fit the wide variety of applications. The inherent relation between developing new magnetic materials with high performance, and the developments in the technological and industrial sectors, can be clearly demonstrated by the ever increasing demands of these sectors for higher and higher efficiencies on the one hand, and the role of newly developed materials in promoting new emerging technologies, on the other hand.

Relatively high metallicity of magnetic materials for practical applications impose limitations on their efficient use due to their unfavorable characteristics, such as high dielectric and magnetic losses, corrosion, and possibly poor mechanical properties. Accordingly, magnetic oxides with ferromagnetic properties emerged as the most widely used magnetic materials for practical applications, owing to their characteristic high resistivity and low eddy current losses, chemical stability, simplicity of production in mass quantities, and other favorable characteristics. An important class of these oxides is the class of hexagonal ferrites developed in the early 1950's, which dominated the world market of permanent magnet applications by the end of the 1980's. Among these ferrites, the magnetoplumbite (M-type) hexaferrite, is produced nowadays in large quantities at very competitive low prices, thus providing the permanent magnet market with probably the most cost-effective magnetic material. In addition, the possibility of tuning the magnetic properties of these materials over a wide range by suitable metallic substitutions for iron have contributed significantly to the numerouse recent advances in technologies such as magnetic recording and microwave applications.

This concise book is intended to provide an overview of the basic concepts of magnetism and magnetic properties pertinent to permanent magnetic materials, importance of these materials in terms of their market share and versatility of practical use, synthesis techniques, and routes adopted for the modification and tuning of their magnetic properties. In the course of presentation, emphasis is made on hexaferrite materials for permanent magnet applications, with M-type ferrites as the focal point. The discussion, however, is brief, in an attempt to provide a wide spectrum of knowledge in a book with limited volume for quick reference to specialized scientists and engineers. While the scope and depth of the discussion is not comprehensive in terms of subject matter or rigor of the theoretical background, the authors-selected material for this book in an attend to give a broad prospective of the subject matters addressed, referring to the most relevant scientific findings, according to authors' point of view, in a small portion of a voluminous literature in the respective fields of study.

Chapter 1 is dedicated to a brief discussion of the basic principles of magnetism pertaining to the subject matter of the book, including the origin of magnetism and magnetic interactions, magnetic anisotropies, and the performance of a permanent magnet in a magnetic circuit. At the end of the chapter, magnetic units are discussed in detail, relying on derivations of the various quantities from the basic electromagnetic equations.

Chapter 2 is concerned with the historical development of permanent magnets, with emphasis on the main magnets in practical use nowadays, namely, alnico, ferrite, and rare-earth permanent magnets. Statistics pertaining to the market share of these materials, and level of interest of both scientists and technologists in the development and improvement of their properties, occupied a good portion of this chapter. Also, some of the basic concepts related to magnet performance are addressed.

Chapter 3 is devoted to the structural and magnetic properties of some hexaferrites, with emphasis placed on M-type hexaferrites. The major routes of synthesis adopted for the preparation of the magnetic ferrite powders are described in this chapter.

Chapter 4 is mainly concerned with the methods of modifying and tuning the magnetic properties of M-type hexaferrites. Here, effects of the synthesis routes and experimental conditions, as well as the role of the various strategies of cationic substitutions are addressed. Emphasis in this chapter was directed toward improving the magnetic properties for hard permanent magnet and microwave applications on the one hand, and for magnetic recording applications on other hand.

Chapter 5 and chapter 6 are devoted to the discussion of some most important applications of hard ferrite magnets, where chapter 6 was devoted to magnetic recording. The separation of the applications into two different chapters is due to the different requirements of magnetic recording, and other types of applications requiring hard permanent magnet properties. In chapter 5, applications in the field of power conversion actuators and transducers, with emphasis on motors and speakers which have the largest market share of permanent magnet applications, were briefly discussed. Also, some of the important passive microwave devices where hexaferrites play an important role in improving device operation were discussed. The discussion of all devices, however, was neither exhaustive in terms of addressing all aspects of a given application, nor complete in terms of addressing all types of devices. Thus these chapters are intended to give the reader a general idea about the importance of ferrite magnets for practical applications.

Sami Mahmood
The University of Jordan, Amman
July 2016

CHAPTER 1

Basics of Magnetism

I.O. Abu Aljarayesh

Physics department, Yarmouk University, Irbid, Jordan

ijaraysh@yu.edu.jo

Abstract

This chapter is dedicated to the discussion of the basic principles related to magnetism. The origin of magnetic dipole moments for free atoms was reviewed. Magnetic ordering and the main classes of magnetic materials were defined. The essential magnetic energy terms (exchange, anisotropy, magnetostatic) were briefly explained. The magnetic domain theory together with the hysteresis loop were also explained. The demagnetizing curve, together with its relevant magnetic parameters were explained. Also, the basic design parameters of a permanent magnet were discussed. Finally, a derivation of the conversion relations of the magnetic units was carried out.

Keywords

Magnetic Dipole Moments, Magnetic Materials, Free Energies of a Ferromagnet, Domains, Hysteresis, Permanent Magnets

Contents

1.   THE ORIGIN OF MAGNETIC MOMENTS

Atoms in a solid may have a net magnetic moment, which results from electrons in the atomic orbitals. The magnetic moment associated with an electron in an orbit is a consequence of the presence of an orbital angular momentum, L, and a spin angular momentum S. Contrary to the classical description, the quantum mechanical treatment of the electronic states demonstrated that angular momentum of an electron can only assume a set of discrete values determined by the corresponding quantum numbers. It is known from quantum mechanics that the state of each electron in a free atom is determined by a set of four quantum numbers [1-7], these are:

1. The principal quantum number (n) which takes the values:

$$n = 1, 2, 3, \ldots \tag{1}$$

This determines the energy of the shell.

2. The orbital angular momentum quantum number (l), which takes the values,

$$l = 0, 1, 2, \ldots, (n\text{-}1) \tag{2}$$

An electron in a given orbit possesses an orbital angular momentum L, with an upper limit of l being $(n\text{-}1)$. The Eigen values of L are given by:

$$L = \sqrt{l(l+1)}\hbar \tag{3}$$

Here $\hbar = h/2\pi = 1.054 \times 10^{-34}$ J.s, where h is Planck's constant.

3. The magnetic quantum number (m_l), which takes the values,

$$m_l = +l, (l-1), \ldots, 0, -1, -2, \ldots, -l \tag{4}$$

Accordingly, there are $(2l + 1)$ different electronic states associated with each shell characterized by a principal quantum number n and an orbital quantum number l. This number of states represents the possible projections of L on the z- direction, resulting in the well-known quantization of space. The z-components of the orbital angular momentum are given by $L_z = \hbar m_l$.

4. The spin quantum number s, is a result of introducing a relativistic correction to the Schrödinger equation, and it has no classical analog. The Eigen values of S are:

$$S = \sqrt{s(s+1)}\hbar \tag{5}$$

An electron is a fermion with spin quantum number $s = \frac{1}{2}$. The projection of S on the z-axis is characterized by the quantum number m_s which according to Eq. (4) takes the values $\pm \frac{1}{2}$. Accordingly, the z-components of the spin angular momentum are given by $S_z = \pm \hbar/2$.

The orbital magnetic moment of an electron with orbital angular momentum L is given by:

$$\mu_l = -(e/2m_e)L \text{ (SI), and } \mu_l = -(e/2m_e c)L \text{(cgs)} \tag{6}$$

where e is the electron charge, and m_e is the mass of the electron. The magnitude of the magnetic moment associated with an orbital quantum number l is therefore given by:

$$\mu_L = \mu_B \sqrt{l(l+1)} \tag{7}$$

where μ_B is a universal constant called Bohr magneton. In SI system of units, Bohr magneton is given by: $\mu_B = e\hbar/2m_e = 9.27 \times 10^{-24}$ J/T or A.m^2. In cgs system, on the other hand, Bohr magneton is given by: $\mu_B = e\hbar/2m_e c = 9.27 \times 10^{-21}$ erg/Oe (or simply, emu).

Also, the magnetic moment associated with the spin angular momentum S is given by:

$$\boldsymbol{\mu_S} = -(e/2m_e).\,2\boldsymbol{S} \text{ (SI), and } \boldsymbol{\mu_s} = -(e/2m_ec).\,2\boldsymbol{S} \text{ (cgs)} \tag{8}$$

The magnitude of the spin magnetic moment of an electron with spin quantum number s is therefore given by:

$$\mu_l = 2\mu_B\sqrt{s(s+1)} \tag{9}$$

The total magnetic moment of an electron is the vector sum of the magnetic moments associated with orbital motion and spin, and is given by:

$$\boldsymbol{\mu_T} = \boldsymbol{\mu_L} + \boldsymbol{\mu_S} = -(e/2m_e)(\boldsymbol{L} + 2\boldsymbol{S}) \text{ (SI)} \tag{10}$$

The total magnetic moment is proportional to the total angular momentum, and is usually expressed as:

$$\boldsymbol{\mu_T} = g\mu_B\boldsymbol{J}/\hbar \tag{11}$$

where the factor g is called the Lande g-factor. Further, the total angular momentum of an electron is given by the vector sum of its orbital and spin angular momenta:

$$\boldsymbol{J} = \boldsymbol{L} + \boldsymbol{S} \tag{12}$$

The Eigen values of the total angular momentum of the electron may be given in terms of the total angular momentum quantum number J by:

$$|\boldsymbol{J}| = \sqrt{J(J+1)}\hbar \tag{13}$$

The z-component of the total angular momentum is $J_z = m_J\hbar$, where m_J assumes values from $-J$ to $+J$, just as suggested by Eq. (4). The g-factor can be evaluated by equating the dot products of the right-hand-sides of Eq. (10) and Eq. (11) with $\boldsymbol{J}$, where we obtain:

$$g\mu_B\boldsymbol{J}.\boldsymbol{J} = \mu_B\boldsymbol{J}.(\boldsymbol{L} + 2\boldsymbol{S}) = \mu_B\boldsymbol{J}.(\boldsymbol{J} + \boldsymbol{S}) \tag{14}$$

Accordingly we obtain the result:

$$g = \frac{\boldsymbol{J}.(\boldsymbol{J}+\boldsymbol{S})}{\boldsymbol{J}\boldsymbol{J}} = 1 + \frac{\boldsymbol{J}\boldsymbol{S}}{\boldsymbol{J}\boldsymbol{J}} = 1 + \frac{\boldsymbol{L}.\boldsymbol{S}+\boldsymbol{S}.\boldsymbol{S}}{\boldsymbol{J}\boldsymbol{J}} \tag{15}$$

Using the identity

$$\boldsymbol{L}.\boldsymbol{S} = \tfrac{1}{2}(\boldsymbol{J}.\boldsymbol{J} - \boldsymbol{L}.\boldsymbol{L} - \boldsymbol{S}.\boldsymbol{S}) \tag{16}$$

in Eq. (15), with the definitions of the magnitudes of the angular momenta in Eq. (3), Eq. (5) and Eq. (13), we obtain the value of g in terms of the angular momentum quantum numbers, where:

$$g = \frac{3}{2} + \frac{[s(s+1) - l(l+1)]}{2J(J+1)} \qquad (17)$$

which is a pure number. For $l = 0$, we have $J = s$, and therefore $g = 2$. Introducing relativistic quantum effects results in a small correction to this value, and $g \approx 2.0023$ [3]. This value was found to be consistent with the experimental values for magnetic transition metals (Fe, Co, and Ni), as well as for several ferromagnetic and paramagnetic alloys and compounds, suggesting that ferromagnetism (parallel alignment of the magnetic moments) in transition metals is due to spin only [8]. This phenomenon is known as *quenching of the orbital moment* [9], which results from the action on the magnetic atom of the crystal field produced by the surrounding atoms, leading to distortions of the electron orbits, and a strong coupling of the orbits to the crystal lattice.

So far, we have established that there is a magnetic dipole moment associated with each electron. When we deal with a free atom of n-electrons, the net orbital angular momentum of an atom is the vector sum of the orbital angular momenta of the constituent electrons. Similarly, the net spin angular momentum of the atom is the vector sum of the spin angular momenta of all electrons. For a completely filled shell, however, the net orbital and spin angular momenta add up to zero, resulting in a zero total angular momentum and the magnetic moment of the atom vanishes. The magnetic moment of an atom, therefore, results from the non-vanishing orbital and spin contributions of electrons in partially filled shells. This is evident in the case of transition metals with partially filled 3d, 4d, 5d and 4f shells, for example.

The total angular momentum of a partially filled shell is calculated within Hund's rules [4-6], where electrons are arranged in the atomic orbitals according to Pauli exclusion principle, and in a manner such that:

Rule one: The total spin angular momentum S is maximum.

Rule two: The total orbital angular momentum L is maximum in accordance with *rule one*.

Rule three: The total angular momentum J is given by:

$$J = \begin{cases} L + S \text{ for more than half} - \text{filled shells} \\ |L - S| \text{ for less than half} - \text{filled shells} \end{cases}$$

For example, the electronic configuration of a free iron atom with 26- electrons is:

$1s^2$) $2s^2$ $2p^6$) $3s^2$ $3p^6$ $3d^6$) $4s^2$

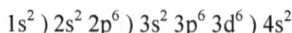

The total angular momentum and net magnetic moment of the atom in this case are determined by the contributions of the 3d unfilled shell with quantum numbers: $n = 3$, and $l = 2$, $m_l = +2, +1, 0, -1, -2$ and $m_s = \pm \frac{1}{2}$. According to Hund's rules, the five orbitals having different m_l values are filled with spin-up ($m_s = +\frac{1}{2}$) electrons in accordance with *rule one*. The sixth electron with spin down occupies the orbital with $m_l = 2$, in accordance *rule two*. This procedure results in the following electron arrangement in the 3d shell:

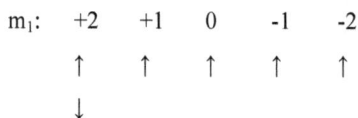

$$m_1: \quad +2 \quad +1 \quad 0 \quad -1 \quad -2$$

$$\uparrow \quad \uparrow \quad \uparrow \quad \uparrow \quad \uparrow$$

$$\downarrow$$

Accordingly, the net spin of the atom is the sum of four parallel spins ($S = 4 \times (1/2) = 2$). Similarly, the net orbital momentum is determined by the contribution of the sixth electron, since the net orbital contribution of the first five electrons adds up to zero; hence, $L = 2$. Further, *Rule three* indicates that $J = L+S = 4$.

Thus, we have established that a net magnetic moment is possessed by a free atom, only if it has a partially filled shell. Of course, there is also a magnetic moment associated with the protons and neutrons of the nucleus, since these also have spin angular momenta. The corresponding magnetic moment, however, is several orders of magnitude smaller than that of an electron due to the large masses of these particles, where the mass of the proton or the neutron is about 1830 times as large as that of the electron. Accordingly, the nuclear magnetic moment is usually neglected in the analysis of the magnetic data.

Magnetism is established in solid materials as a consequence of a complex scenario of interactions of electrons within an atom, as well as the interactions of the electrons of one atom with the electrons of other atoms. Moreover, due to their presence in a crystal, atomic moments interact with one another in ways leading to a variety of magnetic behavior of the different magnetic materials [10, 11].

These interactions in general can modify the magnetic moments of the individual atoms, and can create additional magnetic moments. The degree of modification depends on the type of the material, (metal, insulator, rare- earth compound, alloy, etc.), as well as on the existing condition (temperature, pressure, treatments). In any case, this is a many- body problem, and its theoretical solution is available through some kind of modeling. There exist many models in the literature, the simplest of which is the mean field model introduced by Weiss in 1907 [3, 7, 12]. In general the models can be classified into three main categories: the localized magnetism models, and the itinerant magnetism models, and the so called spin fluctuation models (i.e., the degree of localization) [4, 7].

2. CLASSIFICATION OF MAGNETIC MATERIALS

The magnetic materials are classified into five main kinds; these are: diamagnetic, paramagnetic, ferromagnetic, ferrimagnetic, and antiferromagnetic [2, 8]. It is common to use the response of the material in the linear response region to external fields as a basis of classification of materials. This response is called the susceptibility (χ), which is given by:

$$\chi = \frac{\partial M}{\partial H} \tag{18}$$

Since the main concern of this book is permanent magnets, we will focus our attention in the forthcoming discussion on materials exhibiting substantial spontaneous magnetization, namely, ferromagnets and ferrimagnets. Also, analysis of the magnetic behavior of paramagnets will be addressed in some details, since such analysis is essential for understanding the magnetic behavior of ferromagnetic and ferrimagnetic materials.

2.1 DIAMAGNETISM

Diamagnetic materials are composed of atoms or ions, each with zero magnetic moment. This is a consequence of closed shell electronic configuration, leading to $L = 0$ and $S = 0$ for each individual atom or ion. Diamagnetic response is therefore a consequence of induced magnetism in accordance with Lenz's law. All monoatomic inert gases starting with helium (He) and ending with radon (Rn), as well as most diatomic gases such as N_2 and H_2 are diamagnetic. In addition, solids formed by bonding mechanisms leading to closed shell configurations are also diamagnetic. Accordingly, ionic solids such as alkali halides are diamagnetic. Also, closed shell configurations by covalent bonding are responsible for diamagnetism in diamond, silicon, and germanium, as well as most organic compounds. This kind of magnetism, however, is not limited to these materials; all materials exhibit diamagnetism. The weak nature of this magnetism, however, is masked in materials exhibiting much stronger kinds of magnetism such as paramagnetism and ferromagnetism. The susceptibility of a diamagnetic is very small, negative, and is temperature independent.

2.2 PARAMAGNETISM

The constituent atoms possess permanent magnetic moments, but the orientations of these moments are completely random, due to thermal agitation, and the absence of interactions between the magnetic moments. When an external magnetic field is applied, it exerts a torque (τ) on the magnetic moment that tends to align the magnetic moments partially in the field direction. The magnetic energy (ε) of a magnetic dipole μ in the presence of a uniform magnetic field H applied in the z-direction is given by:

$$\varepsilon = -\mu.H = -g\mu_B H m_J \tag{19}$$

The thermal average of the magnetic moment of an atom in the field direction, using Boltzmann's statistics is:

$$\bar{\mu}_z = \sum_{m_J=-J}^{+J} g\mu_B m_J e^{\frac{g\mu_B m_J H}{k_B T}} \bigg/ \sum_{m_J=-J}^{+J} e^{\frac{g\mu_B m_J H}{k_B T}} \tag{20}$$

For N atoms per unit volume, the average magnetization is given by:

$$M = N\bar{\mu}_z = NgJ\mu_B B_J(x) \tag{21}$$

where $x = g\mu_B J H / k_B T$, and $B_J(x)$ is known as the Brillouin function which is given by:

$$B_J(x) = \left(\frac{2J+1}{2J}\right) \coth\left[\left(\frac{2J+1}{2J}\right)x\right] - \frac{1}{2J}\coth\left(\frac{x}{2J}\right) \tag{22}$$

Since $gJ\mu_B$ is the maximum of each atom in the field direction, the saturation magnetization of the sample would be: $M_0 = NgJ\mu_B$, and the magnetization of the sample given by Eq. (21) reduces to the form:

$$M = M_0 B_J(x) \tag{23}$$

In the classical limit ($J = \infty$), the Brillouin function reduces to the classical Langevin function $L(x)$, and the magnetization is given by:

$$\frac{M}{M_0} = L(x) = \coth x - \frac{1}{x} \tag{24}$$

Practically, magnetic measurements on paramagnets are normally carried out at moderate temperatures and fields ($H \sim 10,000$ Oe). Under these conditions, x in Eq. (21) is rather small, and the arguments of the coth function in Eq. (22) are small. In this case, $\coth y \cong 1/y + y/3$ is a good approximation. Introducing this expansion into Eq. (21), and using the definition of x, we obtain the result:

$$M = NgJ\mu_B \frac{(J+1)}{3J} \frac{g\mu_B J H}{k_B T} = \frac{Ng^2 J(J+1)\mu_B^2 H}{3k_B T} \tag{25}$$

Eq. (25) has the form of the Curie law:

$$\chi = \frac{C}{T} \tag{26}$$

where Curie constant is given by:

$$C = \frac{Ng^2 J(J+1)\mu_B^2}{3k_B} \qquad (27)$$

Thus the susceptibility is small, positive and is inversely proportional to the absolute temperature.

The paramagnetic state is exhibited by the salts of the transition metals, and by the magnetically ordered materials above their critical temperature. At large applied fields and low temperatures, a paramagnetic material can be saturated. The full magnetization curve (magnetization versus applied field strength) can then be compared with the predictions of the quantum and classical theory (Eq. (23) and Eq. (24), respectively). Specifically, the experimental results of the magnetization measurements on potassium chromium alum ($KCr(SO_4)_2.12H_2O$) revealed 99.5% saturation at 1.29 K and field strength of 50,000 Oe [9]. The experimental magnetization curve was in good agreement with the predictions of the quantum theory, Eq. (23), with $g = 2$, which is consistent with $J = S$, and $L = 0$. According to Hund's rules, however, Cr^{3+} ion (the only magnetic ion in the compound) with electronic configuration $1s^2$) $2s^2$ $2p^6$) $3s^2$ $3p^6$ $3d^3$) is associated with $S = 3/2$, $L = 3$, and $J = 3/2$. These values, when substituted in Eq. (17), lead to $g = 2/5$. The calculated magnetization curve using this value of g in the Brillouin function is dramatically different from the experimental results. Further, the experimental magnetization curve was not in satisfactory agreement with the predictions of the classical theory given by Eq. (24) The results of this study, therefore, elegantly confirmed both space quantization, and the quenching of the orbital moments in transition metals.

2.3 FERROMAGNETISM

In a ferromagnetic material, the intrinsic interactions promote parallel alignment of the magnetic moments. The susceptibility of a ferromagnetic material is a function of both magnetic field intensity and temperature, and typically 10^5 greater than that of a paramagnetic material. Ferromagnetic materials lose their magnetic order gradually as the temperature increases up to a specific critical temperature, called Curie temperature (T_c), at which the material undergoes a second order phase transition to a paramagnetic state. At temperatures exceeding this critical transition temperature (in the paramagnetic regime), the temperature dependence of the susceptibility of a ferromagnetic material was found to be consistent with the more general Curie-Weiss law:

$$\chi = \frac{c}{T-T_c} \qquad (28)$$

In an attempt to explain this behavior, Weiss (1907) used the theory of paramagnetism, with the presence of interactions between the individual moments in the solid [12]. These interactions are responsible for the magnetization of the material, which was viewed as a fictitious internal field acting on the individual moments in addition to the applied field (H_a); this field was called the molecular field (H_m). According to Weiss, the internal magnetic field is proportional to the macroscopic magnetization, and is given by:

$$H_m = \lambda M \tag{29}$$

where λ is the molecular field constant, which is a characteristic constant independent of temperature. The effective field acting on the magnetic moment is therefore $H = H_a + H_m$, and the magnetization given by Eq. (21) is modified and becomes:

$$M = N\bar{\mu}_z = NgJ\mu_B B_J \left(\frac{g\mu_B J}{k_B T} (H_a + \lambda M) \right) \tag{30}$$

In the limit of small argument of the Brillouin function, the function is expressed in terms of the first two non-vanishing terms as before. Accordingly, the magnetization is given by:

$$M = \frac{C}{T}(H_a + \lambda M) \tag{31}$$

Here C is given by Eq. (27). The solution of this equation for M is obtained after simple manipulation, and the result is:

$$M = \frac{C}{T - \lambda C} H_a \tag{32}$$

Observing that the susceptibility in the linear region of the Brillouin function is given by M/H_a, we obtain the susceptibility for a ferromagnet in the form:

$$\chi = \frac{C}{T - \lambda C} \tag{33}$$

Comparing this equation with Eq. (28) reveals that the critical transition temperature ($T_c = \lambda C$) is dependent on the molecular field constant, which is, in a way, a measure of the strength of the magnetic interactions within the material. Accordingly, using the definition of C in Eq. (27), the Curie temperature is given by:

$$T_c = \lambda \frac{Ng^2 J(J+1)\mu_B^2}{3k_B} \tag{34}$$

In the ferromagnetic region at $T < T_c$, the material exhibits spontaneous magnetization (M_s), which is ideally equal to M_0 at 0 K due to perfect parallelism of the individual magnetic moments. In the absence of an applied field, the magnetization is given by Eq.

(30) with $H = H_m = \lambda M$. Curve (a) in Fig. 1 shows the magnetization curve as a function of H_m at a constant temperature. On the other hand, the variation of the magnetization with the molecular field is represented by the straight line (b), whose slope is $1/\lambda$. The spontaneous magnetization M_s is determined by the intersection of the straight line with the magnetization curve at point P. It is worth mentioning at this point that the intersection of the straight line with the magnetization curve at the origin represents an unstable state, since a minute magnetization of the sample due to even a very weak stray field results in a non-vanishing molecular field in accordance with Eq. (29). As this molecular field acts, the magnetization moves up along curve (a), resulting in an increase in the magnetization, and consequently, an increase in the molecular field strength. This increase moves the magnetization further up the curve, which finally reaches point P in a zigzag manner. Similarly, if the magnetization exceeds point P, it moves spontaneously down the curve in a zigzag motion back to point P, since the straight line in this region lies above the magnetization curve. Point P, therefore, is a stable magnetic state, demonstrating the presence of a spontaneous magnetization in the sample even in the absence of an external applied field. At the critical (Curie) temperature, the straight line (c) representing the dependence of the molecular field strength on the magnetization becomes tangent to the curve at the origin, and the spontaneous magnetization vanishes, resulting in the transition to the paramagnetic state.

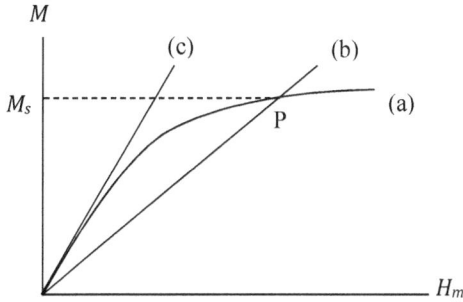

Fig. 1: Development of the spontaneous magnetization by the presence of the molecular field in a ferromagnetic material.

The derivation of the spontaneous magnetization then follows in a straight forward manner. Using the definition of x in the Brillouin function, with $H = H_m$, and after little manipulation we obtain:

$$\frac{M}{M_0} = \frac{k_B T}{\lambda N g^2 \mu_B^2 J^2} x \tag{35}$$

Combining this equation with Eq. (34) we obtain:

$$\frac{M}{M_0} = \frac{J+1}{3J} \frac{T}{T_c} x \tag{36}$$

Also, the magnetization is determined by the Brillouin function, where:

$$\frac{M}{M_0} = \frac{2J+1}{2J} \coth\left(\frac{2J+1}{2J}\right) x - \frac{1}{2J} \coth\left(\frac{1}{2J}\right) x \tag{37}$$

Eq. (36) shows that M/M_0 is linear in x, with a slope which increases with increasing T/T_c. On the other hand, Eq. (37) shows that the relative magnetization is a curve determined by the Brillouin function. The spontaneous magnetization, at any temperature, is then determined graphically from the intersection of the straight line with the curve at point P as illustrated by Fig. 2.

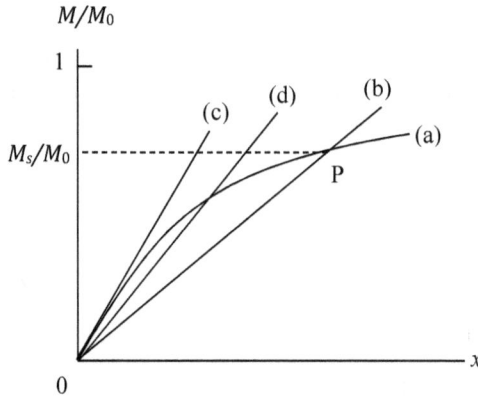

Fig. 2: Plots of Eq. (36) and Eq. (37). The straight lines are drawn for different temperatures, where $T_c > T_d > T_b$.

Notice that at 0 K, the slope of the straight line is zero as inferred by Eq. (36), and the line intersects the curve at $x = \infty$. At this point, Eq. (37) reveals that $M/M_0 = [(2J+1)/2J] - 1/2J = 1$. As the temperature increases, the slope of the line increases, and the point of intersection moves down the Brillouin curve, resulting in a reduction of the spontaneous magnetization. At $T = T_c$, the slope of the straight line is equal to the slope of the Brillouin function at the origin, and the point of intersection occurs at $x = 0$; at this point, $M/M_s = 0$. In the special case of $J = \frac{1}{2}$, where the magnetic moment is determined solely by the spin, with no role played by the orbital motion, Eq. (36) and Eq. (37) reduce to the forms:

$$\frac{M}{M_0} = \frac{T}{T_c} x \tag{38}$$

$$\frac{M}{M_0} = \tanh(x) \tag{39}$$

Combining these equations at the point of intersection we obtain:

$$\frac{M_S}{M_0} = \tanh\left(\frac{M_S/M_0}{T/T_c}\right) \tag{40}$$

This equation can be solved numerically for the dependence of the relative spontaneous magnetization on the relative temperature. Fig. 3 shows the calculated spontaneous magnetization as a function of T/T_c. This curve shows that the spontaneous magnetization decreases only slightly as the temperature increases up to half-way to the Curie temperature, after which the rate of decrease enhanced progressively with increasing temperature up to the Curie temperature.

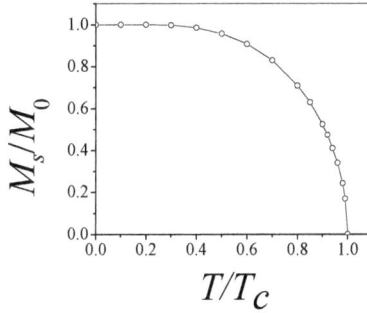

Fig. 3. Numerical solution of Eq. (40) as a function of the relative temperature.

Fig. 4 shows a schematic diagram of the variation of the spontaneous magnetization as a function of T/T_c for $J = \frac{1}{2}$, $J = 1$, and for the classical Langevin behavior with $J = \infty$. Experimental measurements on Fe, Co, and Ni metals were in close agreement with the curve corresponding to $J = \frac{1}{2}$ and $g = 2$. This indicates that the magnetic moment is determined entirely by spin, with no orbital contribution as a consequence of the quenching of the orbital moment as mentioned previously.

At this point, it is worth mentioning that when measurements are carried out at different temperatures, the volume of the sample changes with the temperature, and accordingly, the number density N of magnetic moments changes. Consequently, the magnetizations measured at different temperatures refer to different number densities, inducing an error in the analysis which assumes fixed N values. A more practical measurement is the measurement of the specific magnetization per unit mass ($\sigma = M/\rho$), since the mass of the sample does not change with temperature. The analysis is essentially identical to the analysis of the magnetization, with M replaced by σ in Eq. (36) – Eq. (40).

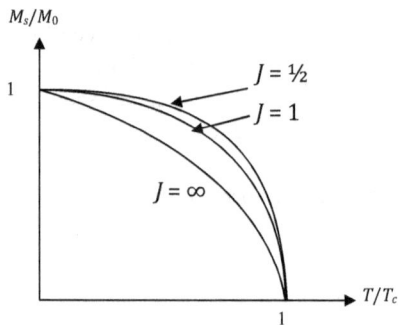

Fig. 4. Schematic variation of the relative spontaneous magnetization with the relative temperature.

2.4   FERRIMAGNETISM

A ferrimagnetic material consists of at least two unequivalent magnetic sublattices, the moments in each of which order ferromagnetically, resulting in a net magnetic moment for each sublattice. Further, the net magnetic moments of the different sublattices are antiparallel, and the fact that the different sublattices have different net magnetic moments leads to a net magnetic moment in the material, and a corresponding spontaneous magnetization as in the case of a ferromagnetic material. In a two-sublattice material such as cubic ferrites $MO.Fe_2O_3$ (M is a divalent metal ion such as Mg, Mn, Ni, Co, Zn), the small metal ions occupy two different crystallographic interstitial sites between the large oxygen anions. These are tetrahedral (A), and octahedral (B) sites. According to Néel [13], the interaction between a magnetic ion on an A site and another on a B site promotes antiparallel alignment of the moments of these ions, which is responsible for the ferrimagnetic structure, and the spontaneous magnetization in ferrites. If the macroscopic magnetizations of the A and B sublattice are respectively $M_A(T)$ and

$M_B(T)$, then the net magnetization of the material is $M(T) = |M_A(T) - M_B(T)|$. Temperature dependent magnetization is the sustained in the material up to the critical transition temperature, also called the Curie temperature, above which the material undergoes a phase transition to the paramagnetic state.

In hexagonal ferrites, the magnetic ions occupy several nonequivalent crystallographic sites. For example, in magnetoplumbite (M-type) hexaferrite, there exist five different interstitial sites for the magnetic ions. These are named: $2a$, $4f_1$, $4f_2$, $2b$, and $12k$. Magnetic interactions between the magnetic moments force the moments on each site to be parallel, forming five different magnetic sublattices. The arrangement of the sublattices in the unit cell, and the relative strengths of the magnetic interactions between ions on neighboring sublattices, promote a unique magnetic structure for this ferrite. The $2a$, $2b$, and $12k$ sublattices align parallel to each other, forming the so called spin-up sublattices, whereas $4f_1$ and $4f_2$ also align parallel to each other, but antiparallel to the previous three sublattices, forming the so called spin-down sublattices. Due to the nonequivalence of the number of magnetic ions in spin-up and spin-down sublattices, the material exhibits non-vanishing spontaneous magnetization below the Curie temperature.

Analysis of the temperature dependent magnetic behavior of a ferrimagnet, based on molecular field hypothesis, reveals that the inverse susceptibility is hyperbolic in temperature, and intersects the temperature axis at the Curie temperature. The high-temperature region of the inverse susceptibility, on the other hand, is linear, just as in the case of a paramagnet. The extrapolated straight, however, intercepts the temperature axis in the negative temperature region, which indicates that the material at high temperatures follow a modified version of the Curie-Weiss law with the sign of the critical transition temperature in Eq. (28) being positive.

2.5   ANTIFERROMAGNETISM

Antiferromagnetism can be considered as a special type of ferrimagnetism, i.e., it consist of two identical sublattices each of which is ferromagnetically ordered, but the magnetic moments of one sublattice are antiparallel to those of the other, and whence, the net magnetization is zero. Antiferromagnetic materials lose their order above a critical temperature called Néel temperature, and transform to the paramagnetic state. Fig. 5 illustrates the susceptibility behavior with temperature for the five kinds of materials discussed above.

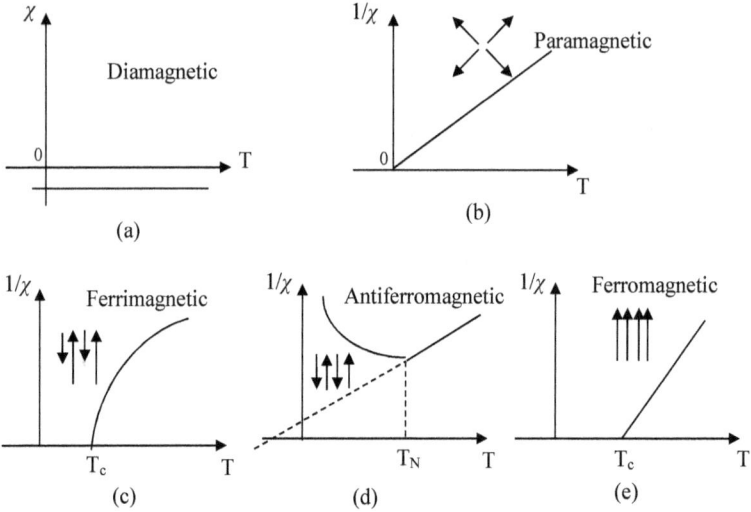

Fig. 5. Schematic representation of the susceptibility behavior with temperature for the different kinds of materials: (a) Diamagnetic, (b) Ideal paramagnetic, (c) Ferrimagnetic, (d) Antiferromagnetic and (e) Ferromagnetic.

3.   EXCHANGE INTERACTIONS

The *exchange interaction* is the essential interaction that is responsible for the observed magnetic order in solids. The origin of this type of interaction stems from the Coulomb interactions between the electrons of the neighboring atoms, taking into account the indistinguishability of the electrons in accordance with Pauli Exclusion Principle [3, 7, 10, 11, 14-17]. But, before introducing an expression for the corresponding exchange energy, let us make some elementary estimation of its magnitude, which may indicate possible sources of this interaction in solids.

In solids, magnetic order is determined by two competing effects, the interactions which are responsible for ordering the magnetic moments, and the randomizing thermal effect. At the critical temperature (which is the temperature above which the material loses its order), the energy of the two interactions is equal. For iron, the critical temperature $T_c \sim$ 1400 K, and the corresponding thermal energy at this temperature $E_{th} \sim k_B T \sim$ 0.1 eV. If we assume that dipole-dipole interaction is the mechanism responsible for magnetic order in iron, the corresponding interaction energy between two dipoles separated by a distance r_{ij} is given by:

$$E_{d-d} \approx \frac{2\mu_0}{4\pi} \frac{\mu_{eff}^2}{r_{ij}^3} \quad \text{(SI)} \tag{41}$$

Here μ_{eff} is the effective magnetic moment of an iron atom ($\mu = 2.2\,\mu_B$ for Fe). At a typical interatomic spacing of ~ 2 Å, $E_{d-d} \sim 10^{-4}$ eV. This energy is too small in comparison with the thermal energy, and therefore cannot be responsible for magnetic ordering at temperatures just below the Curie temperature.

Another possible interaction which could be responsible for magnetic ordering is the spin-orbit coupling (i.e. $\bar{L} \cdot \bar{S}$ coupling). But the orbital angular momentum in 3d-transition metals is quenched, and in the best case scenario of unquenched orbital momentum, estimates indicated that the corresponding interaction energy is less than 10^{-2} eV. This energy is also too small to account for magnetic order just below the Curie temperature of iron.

Yet another interaction which is present in solids is the Coulomb electrostatic interaction between electrons with energy given by:

$$E_c = \frac{1}{4\pi\varepsilon_0} \frac{e^2}{r} \quad \text{(SI)} \tag{42}$$

where ε_0 is the permittivity of vacuum, e is the change of an electron and r is the e - e separation (~ 2 Å). This energy is ~ 7 eV, which should be corrected for the repulsive short-range interaction resulting from Pauli Exclusion Principle. Although this latter interaction reduces the net energy slightly, the corrected energy remains appreciably large in comparison with the thermal energy, and this type of electrostatic interaction is most likely the candidate responsible for magnetic order.

In 1928 both Heisenberg and Dirac independently proposed a Hamiltonian, based on the electrostatic interactions, taking into account Pauli principle, which requires an antisymmetric wave functions for the electrons. The proposed Hamiltonian which takes into account the spin-spin interaction is given by:

$$H = -\sum_{i \neq j} J_{ij}(r_i - r_j) S_i . S_j \tag{43}$$

where $J_{ij}(r_i - r_j)$ is known as the exchange integral, and r_i and r_j are the positions of atom i and atom j, respectively, and S_i and S_j are the total spin of atom i and atom j, respectively. The form of the Hamiltonian in Eq. (43) suggests that the sign of the exchange integral, which minimizes the energy, determines parallelism or antiparallelism in spin ordering. At this point, it is relevant to present some important features of the exchange integral J_{ij}.

The magnitude of $J_{ij}(r_i - r_j)$ represents the degree of overlapping between the electron wave functions. At large interatomic distances, the exchange integral is positive and small. As the atoms get closer, the degree of overlap of electron orbits increases, and the relative strength of J_{ij} increases up to a maximum, and then decreases with a further decreasing the interatomic distances as demonstrated by Bethe-Slater curve in Fig. 6. At small interatomic distances, the sign of the exchange integral becomes negative.

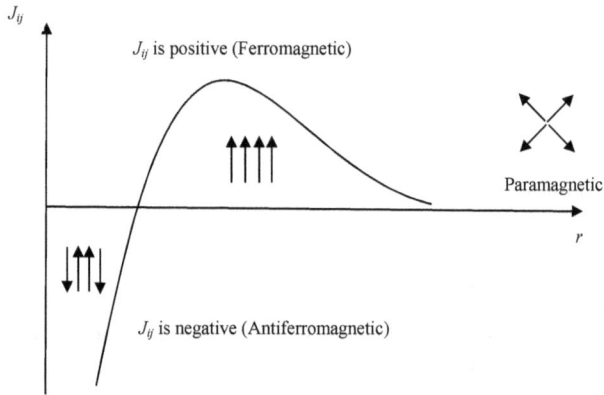

J_{ij}

J_{ij} is positive (Ferromagnetic)

Paramagnetic

r

J_{ij} is negative (Antiferromagnetic)

Fig. 6. Schematic variations of J_{ij} with interatomic distance r_{ij}.

J_{ij} decreases rapidly with increasing the distance between atom i and atom j, since the wave function decays exponentially with distance. This means that the exchange interaction is short-range, and simplification of the problem is achieved by replacing the summation in Eq. (43) by the number of nearest neighbors.

For $J_{ij} > 0$, parallel alignment of the magnetic moments minimizes the energy, and the material is ferromagnetic. If $J_{ij} < 0$, however, antiparallel alignment of the moments minimizes the energy, and the material is antiferromagnetic. At large interatomic distances, the exchange integral vanishes as a consequence of the absence of overlapping between electron orbits, and the exchange energy reduces to zero, leading to a paramagnetic state.

The strength of the exchange interaction determines the value of the critical temperature and the intensity of the so called exchange field.

The Hamiltonian in Eq. (43) is a many-body problem assuming localized moments. Even with the simplification imposed by considering only nearest neighbor interactions, the problem of calculating the exchange energy in magnetic transition metals is not yet solved from first principles. Accordingly, scientists resort to some kind of modeling. In the process, the value of J_{ij} is treated as a parameter to fit the experimental data.

The exchange interaction could be of a *direct exchange* nature, which occurs between the nearest neighbor atoms, usually of the same type, as in the case of $3d$-transition metals. Also, *indirect exchange* interactions between localized moments in some materials are responsible for their magnetic behavior. One of these is the indirect interaction between localized magnetic moments of neighboring atoms via the oscillatory polarization of conduction electrons. This type of interaction, known as RKKY interaction, occurs in rare-earth metals, where the moments of $4f$-electrons in neighboring atoms interact through the polarization of $5s$ and $5p$ conduction electrons [5]. Another indirect type is known as the superexchange interaction, which occurs between two magnetic ions via a nonmagnetic ion, frequently the oxygen anion, as in ferrites. The strength of this interaction is dependent on the interatomic distances (bod lengths), as well as on the M–O–M angle, where M is the magnetic ion and O is the oxygen ion mediating the interaction. For most cases, the exchange integral is negative favoring an antiferromagnetic coupling of the moments [8]. Accordingly, at fixed interatomic distances, the interaction gets stronger as the angle gets closer to 180°, and weaker as the angle gets closer to 90°. This interaction is rather strong in ferrites as reflected by their relatively high Curie temperatures.

Fig. 7 is an illustration of the d-p orbital coupling of two magnetic ions with an oxygen ion in between. The shaded areas represent the overlap between the d-orbitals of the magnetic ions with the p-orbital of oxygen. This is an exchange interaction between the M^{2+} ions through the oxygen ion by sharing an electron of the p-orbital and d-orbital, in accordance with Hund's rules [2].

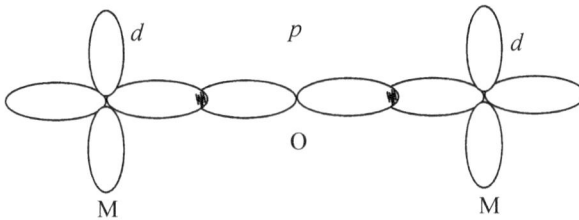

Fig. 7: A schematic representation of superexchange interaction. The angle M–O–M is 180°.

4. MAGNETIC ANISOTROPY

The magnetic anisotropy is the dependence of the free energy of the system on the direction of the spontaneous magnetization (M_s) relative to the crystallographic axes [5, 8, 16-21]. Measurements of the magnetization curves $M(H)$ on single crystals have shown that the magnetization is higher in certain directions than others at the same field, Fig. 8. The shaded area between the two curves represents the magnetic anisotropy energy (MAE). The direction corresponding to the curve with higher magnetization is called the easy axis, while the axes of lower magnetizations are the hard axes.

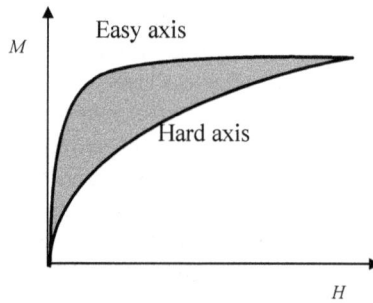

Fig. 8. Magnetization (M) vs. the external field along the easy axis and a hard axis.

The magnetic anisotropy is a decisive factor in determining the details of the hysteresis loop, and thus is essential for any type of applications of ferromagnetic materials. The types of magnetic anisotropies are classified into two main categories: *intrinsic* and *extrinsic* anisotropies. These include *crystal anisotropy, shape anisotropy, stress anisotropy, induced anisotropy* (by magnetic annealing, plastic deformation, or irradiation), and *exchange anisotropy*. Of these, only crystal anisotropy, also known as *magnetocrystalline anisotropy*, is intrinsic to the material, whereas all other types can be modified by synthesis and treatment of the sample. There are several experimental techniques as well as theoretical methods, to estimate the magnetic anisotropy. One of the oldest methods is the area method. For more details, the reader is referred to the discussions of experimental and theoretical methods provided by some authors [8, 22].

In the forthcoming subsections, we discuss the main features of the most important sources of magnetic anisotropy.

4.1 MAGNETOCRYSTALLINE ANISOTROPY

This type of anisotropy arises from the non-spherical charge distribution in the crystal. The physical origin of magnetocrystalline anisotropy is the *spin-orbit coupling*, which might be understood as the interaction between the spin magnetic moment and the orbital magnetic moment. The crystal anisotropy energy is usually expressed as a series expansion in terms of the direction cosines of M_s relative to the crystal axes [5, 8, 23].

For a *uniaxial* hexagonal crystal (Fig. 9), the anisotropy energy depends only on the angle θ between the magnetization vector and the easy c-axis. Further, because of inversion symmetry, only even powers in the expansion give non-vanishing contributions to the crystal energy, which can be expressed as:

$$E_a = \Sigma_{i=0} K_i \sin^{2i}(\theta) \qquad (44)$$

where K_i are the anisotropy constants. Neglecting higher orders [5], the anisotropy energy of a uniaxial is given by:

$$E_a = K_0 + K_1 \sin^2\theta + K_2 \sin^4\theta = K_0 + \sin^2\theta(K_1 + K_2 \sin^2\theta) \qquad (45)$$

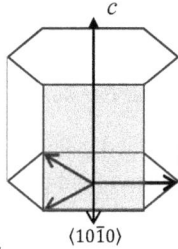

Fig. 9. Hexagonal unit cell.

Eq. (45) indicates that the signs and relative magnitudes of K_1 and K_2 determine the easy magnetization directions in the crystal. Specifically:

1. If K_1 and K_2 are both *positive*, then E_a is minimum at $\theta = 0$, and the c-axis is the easy axis.
2. If K_1 and K_2 are both *negative*, then E_a is minimum at $\theta = 90°$, and thus we have an *easy plane* perpendicular to the c-axis.
3. When K_1 and K_2 have *opposite signs*, then their relative magnitudes define the easy directions. In particular, the following cases arise:

   I. $K_1 > 0; K_2 < 0$

In this case, if K_1 is plotted on the x-axis and K_2 on the y-axis, the line $K_2 = -K_1$ is the boundary between uniaxial and planar anisotropy.

Proof: assuming $K_2 = -\alpha K_1$ (α is positive), the energy is given by:

$$E_a = K_0 + K_1\sin^2\theta(1 - \alpha\sin^2\theta) \tag{46}$$

Thus, if $\alpha < 1$ ($K_2 > - K_1$), the term in parentheses is always positive, and the energy minimum ($E_a = K_0$) occurs at $\theta = 0$, in which case we have an easy c-axis. If, however, $\alpha > 1$ ($K_2 < - K_1$), the term in parentheses could be positive or negative depending on θ. Then the energy is minimum at the most negative value, which occurs at the maximum of $\sin^2\theta$ at $\theta = 90°$, in which case we have an easy plane. When $\alpha = 1$ ($K_2 = - K_1$), both $\theta = 0$ and $\theta = 90°$ are easy directions, since the corresponding energy in each case is minimum with $E_a = K_0$.

II. $K_2 > 0; K_1 < 0$

Assuming $K_1 = - \alpha K_2$, the energy is given by:

$$E_a = K_0 + K_2\sin^2\theta(\sin^2\theta - \alpha) \tag{47}$$

Then,

$$\frac{\partial E_a}{\partial \theta} = 2K_2\sin\theta\cos\theta(2\sin^2\theta - \alpha) \tag{48}$$

$$\frac{\partial^2 E_a}{\partial \theta^2} = 4K_2(3\sin^2\theta\cos^2\theta - \sin^4\theta) - 2K_2\alpha(\cos^2\theta - \sin^2\theta) \tag{49}$$

Here E_a has critical points at: $\theta = 0$; $\theta = 90°$; and $2\sin^2 \theta = \alpha$. However, for $\theta = 0$, the second derivative ($= - 2K_2\alpha$) is always negative, and the energy ($E = K_0$) is maximum in this direction. Thus, we do not have an easy c-axis. In the plane with $\theta = 90°$, the second derivative is:

$$\frac{\partial^2 E_a}{\partial \theta^2} = - 4K_2 + 2K_2\alpha \tag{50}$$

which is positive for $\alpha > 2$. Thus, we have an easy plane with energy $E_a < K_0 - K_2$ when $K_1 < - 2 K_2$, or $K_2 < - K_1/2$.

For $\alpha < 2$ we *do not* have an easy plane (neither an easy axis as explained above). Thus the critical direction is defined by setting the parenthesis in the first derivative equal to zero ($2\sin^2 \theta = \alpha$), which defines a *cone*. In this case, we have:

$$\frac{\partial^2 E}{\partial \theta^2} = 2K_2\alpha(2 - \alpha) \tag{51}$$

Since this is always positive for $\alpha < 2$, we have an *easy cone* of magnetization with half angle $\theta = \sin^{-1}\sqrt{\alpha/2} = \sin^{-1}(\sqrt{|K_1|/2K_2})$. The diagram in Fig. 10 illustrates the easy directions in hexagonal crystals.

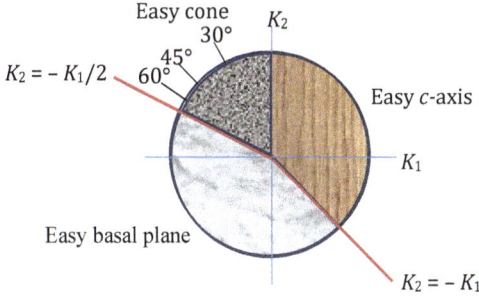

Fig.10. Easy directions in hexagonal crystals for all possible values of K_1 and K_2.

For small rotation of the magnetization direction from the easy direction, the crystal anisotropy acts like a magnetic field trying to rotate the magnetization back to the easy direction. This field is known as the *anisotropy field* (H_K) [8].

Consider an *iron single-crystal* disk cut parallel to the (001) plane and magnetized in the [100] direction. If the direction of M_s makes an angle θ with the [100] easy direction, then $\alpha_1 = \cos\theta$, $\alpha_2 = \cos(90° - \theta) = \sin\theta$, and $\alpha_3 = 0$. The crystal anisotropy energy is then:

$$E_a = K_0 + K_1 \sin^2\theta \cos^2\theta = K_0 + \frac{K_1}{4}\sin^2 2\theta \tag{52}$$

The toque exerted on M_s by the anisotropy field is:

$$\frac{\partial E_a}{\partial \theta} = \frac{K_1}{2}\sin 4\theta = H_K M_s \sin\theta \tag{53}$$

For small θ, we obtain:

$$H_K = \frac{2K_1}{M_s} \quad \text{(cgs)} \tag{54}$$

This formula also applies to *uniaxial crystal* as can be easily verified by considering only the first θ-dependent term in Eq. (45) in the limit of small θ. For a *uniaxial crystal*, the in-plane magnetization curve is given by:

$$H = \frac{2K_1}{M_s}\left(\frac{M}{M_s}\right) + \frac{4K_2}{M_s}\left(\frac{M}{M_s}\right)^3 \tag{55}$$

This is a straight forward result of the minimization of the energy when the field H is applied in the basal plane perpendicular to the easy c-axis. In this case, the field rotates M_s by an angle θ away from the easy axis, and the energy is given by the sum of the anisotropy energy and the magnetic energy due to the applied field:

$$E = K_0 + K_1\sin^2\theta + K_2\sin^4\theta - M_sH\cos(90° - \theta) \tag{56}$$

The condition for energy minimization is then:

$$2K_1\sin\theta\cos\theta + 4K_2\sin^3\theta\cos\theta - M_sH\cos\theta = 0 \tag{57}$$

The in-plane magnetization component produced by the rotation of the spontaneous magnetization vector is:

$$M = M_s\cos(90° - \theta) = M_s\sin\theta \tag{58}$$

Combining Eq. (57) with Eq. (58) by eliminating $\sin\theta$ results in Eq. (55). The in-plane magnetization saturates at a saturation field H_s, which by use of Eq. (55) is given by:

$$H_s = \frac{2K_1 + 4K_2}{M_s} \tag{59}$$

If we set $K_2 = 0$ (which is commonly the case in M-type hexagonal ferrites), the magnetization curve is a straight line:

$$M = \left(\frac{M_s^2}{2K_1}\right)H \tag{60}$$

The saturation field in this case is:

$$H_s = \frac{2K_1}{M_s} \tag{61}$$

This indicates that the anisotropy field for a uniaxial crystal with $K_2 = 0$ is the field required to saturate the crystal in the hard (planar) direction. For BaM hexaferrite at room temperature with $K_1 = 3.3 \times 10^6$ erg/cm^3, $M_s = 72$ emu/g, $\rho = 5.25$ g/cm^3, Eq. (54) gives a value of $H_K \gtrsim 17000$ Oe. Experimental values of half this value are observed. The results of different investigations are usually in poor agreement due to differences between samples imposed by different synthesis techniques. If K_2 is not negligibly small, however, the saturation field is determined by Eq. (59). For hexagonal Co with $M_s = 1422$ emu/cm^3, $K_1 = 4.5 \times 10^6$ erg/cm^3, and $K_2 = 1.5 \times 10^6$ erg/cm^3, the saturation field $H_s \gtrsim 10000$ Oe.

In addition to the strong magnetocrystalline interaction, one may encounter the intrinsic dipole-dipole interaction, with interaction energy given by:

$$E_{d-d} = \sum_{i>j} \left(\frac{\mu_i.\mu_j}{r_{ij}^3} - \frac{3(\mu_i.r_i)(\mu_j.r_j)}{r_{ij}^5} \right) \qquad (62)$$

where μ_i and μ_j are the dipole moments at the positions r_i and r_i, respectively, and r_{ij} is the relative distance between the two dipoles. This is a long-range interaction, in which the dependence of the energy on the relative orientation of the magnetic moments with respect to r_i and r_j, is clear in the second term. In light of the large separations of the moments in ferrites, this energy is seldom considered in the analysis of the magnetic data of these materials.

4.2   SHAPE ANISOTROPY

For polycrystalline materials, the grains are randomly oriented, and the sample as a whole has no crystal anisotropy. In this case, *crystallographic texture* induced by the manufacture of the magnetic material determines the anisotropy, which arises from the weighted average of the crystal orientations exhibiting preference for a certain direction. Different kinds of texture are found in materials depending on the shape of the material body and the manufacturing process. Among these are *fiber texture*, *sheet texture*, *deformation texture*, and *recrystallization texture*. Fiber texture, for example, is often found in round wires, rods, or bar-shaped bodies, where each grain has a preferred orientation of a certain crystallographic direction along the *fiber axis*, which is the wire axis or the rod axis. Sheet texture, on the other hand, occurs in sheets prepared by rolling, where a certain crystallographic plane exhibits preferred parallelism with the sheet surface, and a certain crystallographic direction tends to be parallel to the rolling direction [5, 6, 8].

In a polycrystalline sample with no texture, no crystal anisotropy is observed, such as in the case of a spherical sample where the same filed applied in any direction produces the same magnetization. In a magnetized sample, an internal demagnetizing field ($H_d = -N_dM$; N_d the demagnetizing factor) is induced in an antiparallel direction to the magnetization. Due to the fact that the demagnetizing factor in the direction of a short axis is greater than in the direction of a long axis, a non-spherical sample is easier to magnetize along the long axis of the sample, resulting in anisotropy exclusively determined by the shape of the sample. The demagnetizing field, being in opposite direction to the magnetization, results in a reduction of the magnetization of the sample. Fig. 11 below shows the magnetization curve for the sample as an applied field takes the magnetization to point A, and then turned off. This process induces magnetization in a sample made of a hard magnetic material. The action of the demagnetizing field, however, shifts the magnetization at zero applied field from point R to point C. The magnetic induction in the sample (in cgs units) is given by: $B = H + M = -H_d + M$. The magnetostatic energy stored in a magnetized sample is then given by [8]:

$$E_{ms} = \int_0^M H dM \quad \text{(cgs)} \tag{63}$$

This integral is difficult to evaluate for a sample with a general shape, since the magnetization is generally not uniform in the sample (it is uniform only for an ellipsoid).

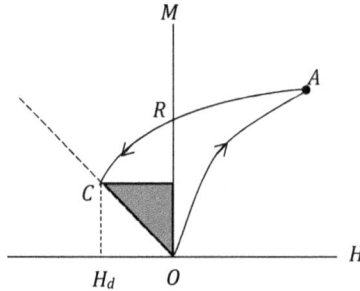

Fig. 11. Magnetostatic energy in a magnetized sample in zero applied field

This energy is positive, since H_d is opposite to the direction of the magnetization, and is given by the shaded area of the triangle in Fig. 11.

$$E_{ms} = \frac{1}{2} H_d M = \frac{1}{2} N_d M^2 \tag{64}$$

Considering a prolate spheroid with $a = b < c$, the energy in terms of the components of M along c, and along a we obtain:

$$E_{ms} = \frac{1}{2}[(M\cos\theta)^2 N_c + (M\sin\theta)^2 N_a] = \frac{1}{2}M^2 N_c + \frac{1}{2}(N_a - N_c)M^2\sin^2\theta \tag{65}$$

This magnetostatic energy has the same angular dependence as the uniaxial crystal anisotropy. The shape anisotropy constant is given by:

$$K_s = \frac{1}{2}(N_a - N_c)M^2 \quad \text{(cgs)} \tag{66}$$

For Co prolate spheroid, K_s approaches the value of $(1/2) \times (4\pi/2) \times M^2 = 63.5 \times 10^5$ erg/cm^3 as the ratio c/a increases. Notice that this is greater than the anisotropy constant for a uniaxial Co crystal.

4.3 INDUCED ANISOTROPY

The induced anisotropy can be created in any process in which the original symmetry is broken, either by processing or treatments, the main types are, Magnetic annealing, in which the material is heated in the presence of a large magnetic field, and thus create a uniaxial material. Also, magnetic anisotropy can be created by creating defects (of many kinds) in the crystal, or by irradiation [5, 8].

There are also, other sources of magnetic anisotropy like magnetostrictions. For Hexaferrites this is small, and usually neglected. A good discussion of these kinds was provided by Cullity and Graham [8].

5.   MAGNETIC DOMAINS

In an attempt to explain the absence of spontaneous magnetization in piece of a ferromagnetic material below its Curie temperature, P. Weiss [2, 7, 24, 25], made an assumption that the sample in the demagnetized state is divided into small regions, called *magnetic domains*. This division is essential for energy minimization. Each magnetic domain is spontaneously magnetized by the strong molecular field, but the directions of magnetization of the individual domains create flux closure (Fig. 12) so that the net magnetic moment of all the domains adds up to zero. Adjacent domains are separated from each other by layers called *domain walls*. The magnetic moments in the layer change their direction gradually from the magnetization direction of one domain toward that of the neighboring domain across the wall. Fig. 13 shows a schematic diagram of the orientations of magnetic moments within a domain wall separating two domains in a uniaxial crystal.

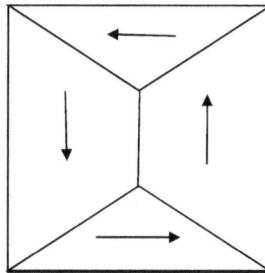

Fig. 12. Schematic representation of magnetic domains in a magnetic material.

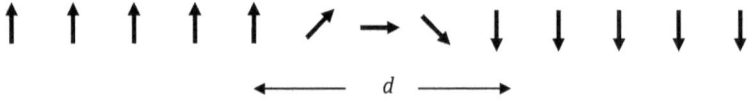

Fig. 13. Schematic representation of a line of moments across a domain wall.

It is relevant to mention that in real materials there is a distribution of domain sizes. Moreover, the structure of these domains, and their walls, together with the dynamics of the wall motion, determine the details of the magnetization curve, which is crucial for determining the suitability of the magnetic material for a particular type of application [25, 26]. In fact, the domain wall dynamics is tailored to maximize certain characteristic to suite a particular application. For example, to increase the coercivity, the domain wall is designed to contain defects, dislocations, wall pining centers, etc. [19].

The domain wall thickness (d), in terms of the number (N) of magnetic moments forming the wall and the distance (a) between adjacent moments, is given by:

$$d = (N - 1)a \sim Na \tag{67}$$

The magnetic structure of a magnetically ordered material is determined by the condition of minimization of the magnetic energy. The magneto static energy E_{mag} of a domain is given by [8]:

$$E_{mag} = \frac{1}{8\pi} \int H^2 \, dv \quad \text{(cgs)} \tag{68}$$

where H is the magnetic field intensity, and the integration is carried over the whole space. Since this energy is practically quadratic in magnetization, the energy of a domain is reduced by splitting the domain into smaller ones. In the process, however, domain walls are created, in which additional energy is stored due to the rotations of the moments within the wall against the crystal anisotropy. The process of division continues, till there is a balance reached between the decrease of magneto static energy and increase of energy needed to create the walls [24, 25].

To estimate the energy of the wall, it is sufficient to consider the main two competing terms in the free energy, namely, the exchange energy (E_{exch}) and the magnetic anisotropy energy (E_a). The exchange energy for two neighboring atoms with the same spin is estimated with [8]:

$$E_{exch} \approx -2J\boldsymbol{S}_i.\boldsymbol{S}_j = -2JS^2\cos\theta_{ij} \tag{69}$$

The angle changes by π across a *Bloch wall* in a uniaxial crystal, which normally contains a large number (N) of moments, leading to a small $\theta_{ij} = \pi/N$. Accordingly, we can use the expansion $\cos\theta_{ij} = 1 - \theta_{ij}^2/2$. Neglecting the term independent of the angle, the exchange energy throughout a line of spins within the wall is then estimated with:

$$E_{exch} \approx JS^2\theta_{ij}^2 = NJS^2\left(\frac{\pi}{N}\right)^2 = JS^2\frac{\pi^2}{N} \tag{70}$$

Since we have $\sim 1/a^2$ lines of atoms per unit area, the total exchange energy of the wall is:

$$E_{exch} \approx JS^2\frac{\pi^2}{Na^2} \tag{71}$$

On the other hand, the magnetocrystalline anisotropy energy per unit area of the wall is KNa, where K is the anisotropy constant. Thus, the total energy is given by:

$$E_T \approx JS^2\frac{\pi^2}{Na^2} + KNa \tag{72}$$

To find N that minimizes the energy of the wall, we require $(\partial E/\partial N) = 0$, and $(\partial^2 E/\partial N^2) > 0$. From these conditions we obtain the relation:

$$E_T = 2\sqrt{\frac{JS^2\pi^2 K}{a}} \tag{73}$$

Notice that $\dfrac{\partial^2 E}{\partial N^2} = 2\dfrac{JS^2\pi^2}{N^3 a^2} > 0.$

And according to Eq. (67), the wall thickness (d) is:

$$d = \sqrt{\frac{JS^2\pi^2}{Ka}} \tag{74}$$

Thus the domain wall thickness depends on the intrinsic properties and structure of the materials. From Eq. (71), it is clear that the exchange energy is lower for a thicker wall, while anisotropy energy tends to increase with the thickness of the wall. A typical wall thickness in hexagonal ferrites is $d \sim 100$ Å.

6. HYSTERESIS LOOP

One of the main characteristics of a ferromagnetic material is its magnetic hysteresis [27, 28]. A hysteresis loop consists of the set of magnetization (M) data points against magnetic field intensity, H, or the corresponding magnetic flux, B against H. These two representations can be derived from each other, using the material relation:

$$B = H + 4\pi M \quad \text{(cgs)} \qquad B = \mu_0(H + M) \quad \text{(SI)} \tag{75}$$

For simplicity, the hysteresis loop of a polycrystalline sample can be sectioned into seven magnetization curves (Fig. 14). These are: OS, which is the initial curve I, SR, RC, CS', $S'R'$, $R'C'$, and $C'S$. The following discussion explains the magnetization processes along each of these curves.

Magnetization of the demagnetized sample starts from the point O, where the moments of the domains are randomly oriented, and the net moment is zero. The magnetic state of the sample at point O is represented by the schematic diagram of the vector arrangement in Fig. 15a, where the vector directions represent the directions of magnetization of the domains in the polycrystalline sample. The magnetization along the OS initial curve occurs at different stages. Along the first part labeled 1 in the figure, the applied magnetic field is low, and the main mechanism of magnetizing the sample is the domain wall motion in the low-field *Rayleigh region*. This region is characterized by a parabolic behavior. At very low fields in the Rayleigh region, however, the curve is linear, due to the weakness of the quadratic contribution, and the magnetization process is reversible. At higher fields, the magnetization is irreversible due to domain wall motion at 180° flipping the magnetization direction from the $- H$ direction to the $+ H$ direction, and hysteresis is observed [8]. As the applied field is increased beyond the maximum value of the Rayleigh region, domains magnetize in directions farther and farther from the $- H$ direction and flip at 180° into directions around the $+ H$ direction in the forward hemisphere. The process continues until a magnetic state similar to that described by Fig. 15b is reached at the point labeled by 2 on the initial magnetization curve. In this illustration of the magnetic state, thicker vectors are used to represent magnetization directions where the 180° flip occurred. As the applied field is increased further, a magnetic state illustrated by Fig. 15c is reached at the point labeled by 3, where the magnetization direction of all domains is now along easy directions in the forward hemisphere of the sample. Up to this point, the magnetization occurs via irreversible domain wall motion. Beyond this point in the high field region, magnetization occurs by virtue of magnetization rotation into directions closer to the direction of the applied field, until a magnetic state similar to that of Fig. 15d is reached at point S. The magnetization rotation processes are reversible due to the competition between the torques of the applied field and the magnetocrystalline anisotropy field.

At point S, the magnetization of the sample is not saturated, and ideally, an infinite field is required to saturate a sample made of a hard magnetic material. In the high field region, the magnetization changes slightly with increasing the applied field intensity, and the magnetization behavior is described by the law of approach to saturation, which is given by:

$$M = M_s \left(1 - \frac{A}{H} - \frac{B}{H^2}\right) + \chi H \qquad (76)$$

jwhere A and B are constants, to be determined from the experimental results. The constant A is associated with inclusions and stress, and the constant B is associated with the magnetocrystalline anisotropy. The last term (χH) is called the *forced magnetization* term, representing the induced increase of the spontaneous magnetization of the domains at high applied fields. At temperatures well below the Curie temperature, this term is negligible as demonstrated by fitting the experimental data for a variety of hexaferrite materials [29-36]. Further increase of the magnetic field, may cause changes in the dimensions of the sample, and thus creates additional anisotropy term arising from *magnetostriction* [8].

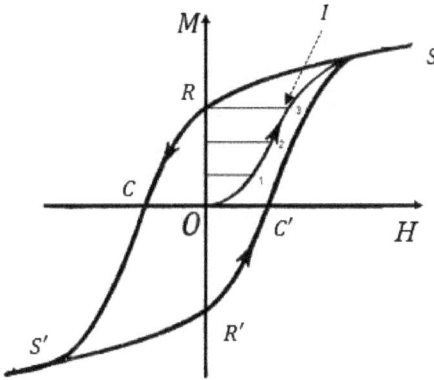

Fig. 14: Hysteresis loop of a typical ferromagnet or ferrimagnet.

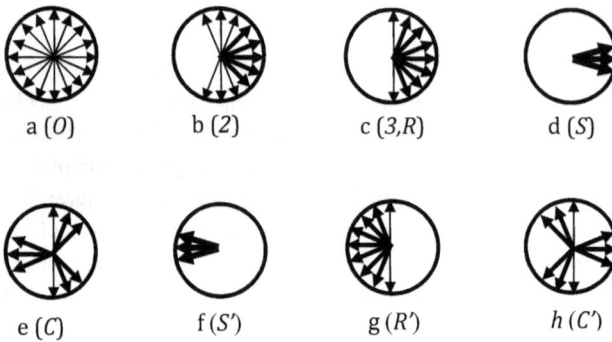

a (O)   b (2)   c (3,R)   d (S)

e (C)   f (S')   g (R')   h (C')

Fig. 15: Schematic diagram of the magnetic states at different points in the initial curve and the hysteresis loop of Fig. 14. (a) represents the magnetic state at O, (b) at point 2 on the initial curve, (c) at point 3 on the initial curve, and point R on the hysteresis loop, (d) at point S, (e) at point C, (f) at point S', (g) at point R', and (h) at point C'.

As the field is decreased, the magnetization follows the path *SR*, exhibiting hysteresis. This happens because as the magnetic field decreases, the magnetization directions of the domains are progressively freed from their rotated positions, until the applied field is zero (at point *R*), when no external torque acts on the magnetization, and all domains become magnetized in their respective easy directions in the forward hemisphere as illustrated by Fig. 15c. The sample in this magnetic state retains a net magnetic moment in the forward direction, and the corresponding magnetization at point *R* is known as the *remanent magnetization*, M_r. The value of M_r is a characteristic one, and reflects the new distribution of magnetization directions in the sample. This value is an important magnetic parameter of the material to be used for practical applications, where relatively high values are required for magnetic recording and permanent magnet applications, for example.

When a reverse field is applied to the sample in the magnetic state defined by point *R*, spin flip by 180° wall motion first occurs, changing the magnetization direction of domains initially with easy axes parallel to +*H* into easy directions in the –*H* direction, resulting in a decrease of the magnetization. As the field is increased in the opposite direction, the magnetization is reduced with increasing the field value till it reaches zero at point *C*. Although the magnetization at this point is equal to that of the initial demagnetized state (*O*), the magnetic state is significantly different as illustrated by Fig. 15e. The field intensity at this point is another characteristic magnetic parameter, called the *intrinsic coercivity* (H_{ci}), which is an important parameter for practical applications. In contrast, if the *B-H* curve is plotted, the field intensity (H_{cB}) at which the induction within the magnet vanishes is often called the *coercivity*. This term, however, is sometimes used by some authors to refer to the intrinsic coercivity, and distinction between the two values is understood from the context. While these two coercivities are almost equal for soft magnetic materials, they could be substantially different for hard permanent magnets, with H_{ci} being higher than H_{cB}.

It is relevant to mention that the behavior of the R $\rightarrow$ C magnetization curve (known as the demagnetization curve) is very crucial for any permanent magnet design. Theoretically, the coercive field is given by an expression derived by Brown [37], assuming a homogenous, uniformly magnetized ellipsoid and is given by [2]:

$$H_c \geq \frac{2K}{M_s} - NM_s \quad \text{(cgs)} \tag{77}$$

where the first term is the value of the anisotropy field for a uniaxial sample, and the second is the demagnetizing field. In real materials a lower value (by 20 – 30%) is usually observed, since real materials are inhomogeneous, and these inhomogeneities are sources for demagnetization effects [2].

Further increase of the field in the negative direction results in an increase of the magnetization in the negative direction, first by domain wall motion and magnetization flip to easy directions in the backward hemisphere, and then by magnetization rotations toward the – H direction, till a state represented by Fig. 15f is reached at point S'. Saturation in the reverse direction can be obtained by further increasing the reverse field intensity.

As the reverse field is reduced, the behavior of $M(H)$ along $S'R'$ is similar to that along the SR curve, where at zero applied field, a magnetic state represented by Fig. 15g is reached, which is essentially the same as that at point R, but reversed. Also, the behavior along $R'C'$ segment is an image of the demagnetization curve RC, and the magnitude of the remanent magnetization in the reverse direction at point R' is equal to that at point R. The magnetic state at C' is represented by Fig. 15h, which is essentially the same as the magnetic state at point C, but reversed. Further, the magnetization processes along $C'S$ are the same as those along CS', where domain wall motion flips the magnetization directions into easy directions in the forward hemisphere first, and then magnetization rotation is responsible for the further increase in magnetization by rotating the spontaneous magnetization directions of the domains toward the direction of the forward applied field. Magnetic saturation can be achieved by increasing the reverse field strength beyond point S'.

The closed loop $SRC\ S'R'C'S'$ is called the hysteresis loop. The area of the hysteresis loop represents the heat dissipated in the material in one cycle, and is equal to the work done on the system according to the first law of thermodynamics.

7.   MAGNET DESIGN

The integral form of one of Maxwell's equations of magnetostatics is given by:

$$\int \boldsymbol{B}.\,d\boldsymbol{a} = 0 \tag{78}$$

This is a restatement of the vanishing of the divergence of the magnetic flux. The integral is evaluated over a closed surface in space. And from Ampere's circuital law in the absence of current sources we have:

$$\oint \boldsymbol{H}.\,d\boldsymbol{l} = 0 \tag{79}$$

The integral here is evaluated along a closed loop. Further, the relation for the magnetic flux density is given by Eq. (75). These equations are essential for the analysis of a magnetic circuit, and thus are crucial for permanent magnet design [38].

Consider an open magnetic circuit made of a permanent magnet ring, from which a section of length l_g has been removed to provide a gap, as shown in the Fig. 16. The permanent magnet produces a magnetic field in the space of the gap, which can be used for device design, such as placing a moving coil in the gap. The length of the gap is l_g and its surface area is A_g. The length of the permanent magnet is l_m and its cross sectional area is A_m. In the absence of fringing of the flux lines in the gap, which is ideally not the case in practice, we have $A_g = A_m$. The flux lines are continuous in the magnet and through the gap, and are given by B_m and B_g, respectively. The H_g-field lines within the gap, however, are in opposite direction to the field lines H_m in the magnet. Consequently, applying Ampere's circuital law to the closed ring circuit leads to the result:

$$H_g l_g - H_m l_m = 0 \tag{80}$$

Accordingly, the field within the gap is given in terms of the field in the magnet by:

$$H_g = \left(\frac{l_m}{l_g}\right) H_m \tag{81}$$

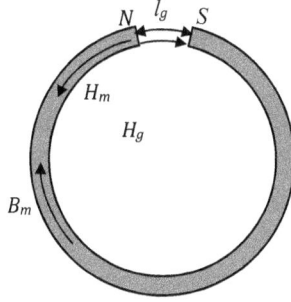

Fig. 16. Illustration of an open magnetic circuit showing the magnetic poles (N and S) and the acting magnetic fields.

Also, from Eq. (78) we obtain another relation. Observing that the integral over the outer surfaces of the ring vanishes, the only contributions to the integral come from integrating across the cross-sectional areas of the gap and magnet. Also, in light of the continuity of the B-lines, and the fact that the direction of the area of the gap is opposite to that of the magnet, we obtain from evaluating the integral in Eq. (78) the following result:

$$B_g A_g = H_g A_g = B_m A_m \quad \text{(cgs)} \tag{82}$$

From this result, we obtain for the gap field intensity:

$$H_g = \left(\frac{A_m}{A_g}\right) B_m \tag{83}$$

Multiplying Eq. (83) by Eq. (81) gives the result:

$$H_g^2 = \left(\frac{l_m}{l_g}\right) H_m \cdot \left(\frac{A_m}{A_g}\right) B_m = \frac{V_m}{V_g}(B_m H_m) \quad \text{(cgs)} \tag{84}$$

This relation demonstrates that the volume of the magnetic material V_m required to produce a given field intensity within a given gap is a minimum at the maximum of $B_m H_m$, i.e., at $(BH)_{max}$. Since the energy per unit volume stored in the field in the air gap is $H^2/8\pi$ (cgs) [8], which is proportional to the square of the magnetic field intensity in the gap, the quantity $(BH)_{max}$ is named the *maximum energy product*. This is an important quality assessment parameter of a permanent magnet constructed for practical applications.

The free magnetic poles in the open circuit create a demagnetizing field within the magnet ($H_d = -N_d M$, where N_d is the demagnetizing factor), lowering the induction below the remanent value B_r (also called *Retentivity* or *residual induction*) for a closed ring. This is inevitable since a permanent magnet is always constructed in an open circuit design to provide a magnetic field in a given space. The level of decrease of the induction depends on the value of the demagnetizing field at the *operating point* of the magnet, which in turn is determined by the magnet design. This is illustrated in Fig. 17, which shows the variation of the induction with magnetic field in the second quadrant demagnetization curve. In the absence of external applied fields, the only acting field is the demagnetizing field in the open magnet circuit. The value of BH determined by the area of the shaded rectangle in Fig. 17 thus depends on the location of the operating point on the demagnetization curve, which can be moved along the curve by design. This is due to the fact that, according to Eq. 2.75), the relation between the induction and the demagnetizing field is expressed in cgs unit as:

$$B = -\frac{4\pi - N_d}{N_d} H_d \tag{85}$$

This straight line with negative slope (since $N_d < 4\pi$) is known as the *load line*. The slope of the load line is determined by the demagnetizing factor N_d, and therefore, the operating point (P) of the magnet as determined by the intersection of load line with the demagnetization curve can be shifted along the demagnetization curve by changing the magnet geometry [8, 39]. By doing so, the energy product of the magnet can be maximized by the designer.

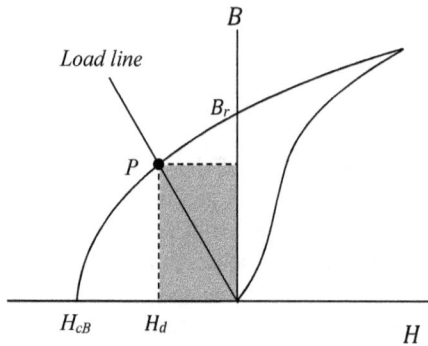

Fig. 17. Schematic graph of the demagnetization Curve, showing the load line and the working point of a permanent magnet.

The shape of the demagnetization curve in the second quadrant is an indicator of the usefulness of the permanent magnet for practical applications as mentioned previously, and is a determining factor for the maximum energy product of the magnet. If the magnetization is flat far into the second quadrant as illustrated by the M1 curve in Fig. 18, the remanent magnetization M_r is equal to the saturation magnetization, and the magnetization is constant up to the intrinsic coercive field value for an ideal rectangular magnetization curve. Accordingly, the induction in this region is a straight line (B1) in accordance with Eq. (75), with an intercept with the B-axis at $B_r = 4\pi M_s$ (cgs) and an intercept with the $-H$-axis at $H_{cB1} = 4\pi M_s = B_r$ as demonstrated by Fig. 18. If the magnetization curve is not so flat as demonstrated by curve M2, the B-H curve (B2) will exhibit a *knee*, and consequently a reduction of the coercivity down to the H_{cB2} value determined by the lower magnetization at this field, even if remanent magnetization and intrinsic coercivity were the same as those for the hypothetical ideal magnet. This is usually accompanied by a reduction of the maximum energy product in comparison with the ideal magnet.

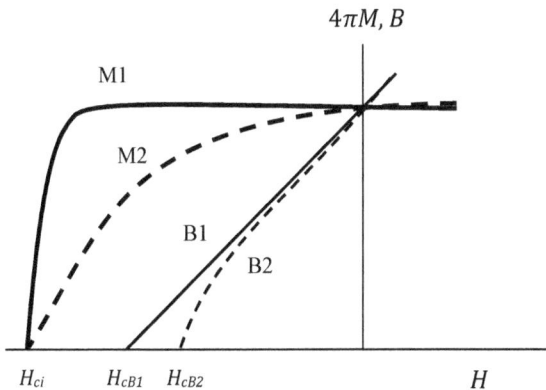

Fig. 18. Magnetization ($4\pi M$) and induction (B) curves for a magnet with almost square M-H curve (M1 and B1), and for a lower quality magnet with a not so flat magnetization curve in the second quadrant (M2 and B2).

For an ideal magnet with a flat magnetization curve far into the second quadrant, the magnetization is constant up to the coercive field. If the linear relation for the induction is multiplied by the magnitude of the field intensity, we obtain the relation:

$$BH = (4\pi M_s - H)H \tag{86}$$

The condition for a critical point in the BH-H relation is determined from the value of H at which the first derivative of the energy product with respect to H is zero. Since the second derivative of the energy product is clearly negative, the critical point is a point of maximum energy product. The condition for a maximum energy product is therefore:

$$H = \frac{4\pi M_s}{2} \quad (cgs) \tag{87}$$

Accordingly, the maximum energy product determined from Eq. (86) is given by:

$$(BH)_{max} = \frac{(4\pi M_s)^2}{4} = \frac{B_r^2}{4} \quad (cgs) \tag{88}$$

Similar analysis reveals that the maximum energy product in the SI system of units is given by:

$$(BH)_{max} = \frac{\mu_0 M_s^2}{4} = \frac{B_r^2}{4\mu_0} \quad (SI) \tag{89}$$

In the preceding discussion, practical problems associated with fringing of the field lines and flux leakage had been ignored. In practice, the right hand sides of Eq. (81) and Eq. (83) should be multiplied by the factors K_1 and K_2, respectively, to account for these effects. Then Eq. (84) should be replaced by:

$$H_g^2 = \frac{V_m}{V_g}(B_m H_m)K_1 K_2 \tag{90}$$

While the effect of fringing is relatively small ($K_1 = 0.7 - 0.95$), the effect of flux leakage could be large ($K_2 = 0.1 - 0.8$). Accordingly, the maximum energy product of the magnet is less than expected, and in the design stages, these effects should be taken into consideration.

8.   MAGNETIC UNITS

The magnetic quantities were traditionally (and commonly nowadays) expressed in *cgs* (*electromagnetic*) units. Recently, the *SI* system of units is adopted, and the demand for converting the magnetic quantities to this system of units is increasing. The *SI* (*rationalized MKSA*) system uses the meter, kilogram, second, and ampere as units for length, mass, time, and current, respectively. The *cgs* system uses centimeter, gram and second as units for the first three quantities. However, the current was given different units in four different fundamental *cgs* systems. The *electromagnetic* (*emu*) and the *Gaussian cgs* systems are perhaps the most common systems used in literature when dealing with electromagnetic phenomena. In the *Gaussian*, as well as in the *electrostatic* (*esu*) system of units, the charge (and consequently the current) is expressed in *electrostatic* (*esu*) units, while in the *emu* system these quantities are expressed in absolute units (*abcoulomb* and *abampere*). The remaining of this section is devoted to deriving the conversion relations for the magnetic quantities between the *emu* (*cgs*) and the *SI* systems of units.

8.1   CHARGE, CURRENT, AND MAGNETIC FIELD

The relation between the *esu* and the *SI* units of charge is derived from Coulomb's law:

$$F = \frac{q\dot{q}}{4\pi\varepsilon_0 r^2} = 9 \times 10^9 \frac{q\dot{q}}{r^2} \qquad (SI) \tag{91}$$

$$F = \frac{q\dot{q}}{r^2} \qquad (esu) \tag{92}$$

According to Eq. (92), the force between two unit charges (1 C each) placed a distance 1 m apart is equal to 9×10^9 N. The numerical value of 1 C in the *esu* system is then determined from Eq. 91:

$$q^2 = F(dynes).r^2(cm^4) = 9 \times 10^9 \times 10^5 \times 10^4 = 9 \times 10^{18} \ (esu^2)$$

$$\Rightarrow q = 1C = 3 \times 10^9 \ esu \tag{93}$$

The conversion factor in Eq. (93) is the same for current.

Now let us consider the magnetic field produced by a current-carrying long straight wire at a normal distance r from the wire. In *electromagnetic* and *Gaussian* systems of units, one of Maxwell equations (Ampere's law) reads:

$$\nabla \times \vec{H} = 4\pi\vec{J} \qquad (emu\ system) \tag{94}$$

$$\nabla \times \vec{H} = \frac{4\pi}{c}\vec{J} \qquad (Gaussian\ system) \tag{95}$$

Here c is the speed of light ($\approx 3\times10^{10}$ cm/s). In both *emu* and *Gaussian* systems, the magnetic field H is expressed in units of *Oersteds* (Oe). Following Michael Faraday's representation of the field strength in terms of *"lines of force"* (1 *line of force* = 1 *maxwell* = 1 *Mx*), the field strength is defined as *the number of lines of force passing through a unit area normal to the field*. Therefore:

1 Oe = 1 line of force/cm^2 = 1 Mx/cm^2

From Eq. (94), the magnetic field is obtained by integrating over the area of a circle whose axis coincides with the wire, and the result is given by:

$$H = \frac{2i}{r} \qquad (Oe) \qquad (emu\ system) \tag{96}$$

However, according to Eq. (95), the field is given by:

$$H = \frac{2i}{cr} \quad (Oe) \qquad (Gaussian\ system) \tag{97}$$

Now according to Eq. (96), a field intensity of 2 Oe (= 2 Mx/cm^2) is generated by a current of 1 abampere at a distance of 1 cm from the wire. The value of this current in esu is given by Eq. (97):

$$i = \frac{crH}{2} = \frac{3\times10^{10}\left(\frac{cm}{s}\right)\times1(cm)\times2(Oe)}{2} = 3 \times 10^{10} \quad (esu) \tag{98}$$

Accordingly, with the use of the conversion factor in Eq. (93) we obtain:

$$1 abampere = 3 \times 10^{10} \quad (esu) = 10\ amperes \tag{99}$$

This last relation provides the numerical conversion relations between the emu, the $Gassian$, and the SI systems for current.

In the SI system, however, Ampere's law is given by:

$$\nabla \times \vec{H} = \vec{J} \qquad (SI\ system) \tag{100}$$

Accordingly, the magnetic field resulting from the long current-carrying straight wire is given is given by:

$$H = \frac{i}{2\pi r} \quad \left(\frac{A}{m}\right) \tag{101}$$

Consequently, a long straight wire carrying a current of 1A would produce a field intensity of $1/2\pi$ (A/m) at a normal distance of 1 m from the wire. In cgs system, according to Eq. (96), this field is:

$$H = \frac{1}{2\pi}\left(\frac{A}{m}\right) = \frac{2 \times \left(\frac{1}{10}\right)}{10^2} \ (Oe) = \frac{2}{10^3} \ (Oe)$$

We therefore have the conversion relation:

$$1\ Oe = \frac{10^3}{4\pi} \frac{A}{m} = 79.6 \frac{A}{m} \tag{102}$$

8.2 MAGNETIC MOMENT

The magnetic moment of a current-carrying loop is the product of the loop area and the current, and is normal to the area. If the area is 1 m^2 and the current is 1 A, the magnetic moment of the loop is 1 A.m^2 in rationalized MKSA units. In cgs electromagnetic system, the magnetic moment of the loop is

$$m = 1A.m^2 = \frac{1}{10}abampere.\,10^4cm^2 = 10^3abampere.\,cm^2$$

The unit of the magnetic moment in the *electromagnetic* system is often called *emu*, even though this abbreviation generally refers to the *electromagnetic* system of units. Therefore:

$$1A.m^2 = 10^3emu \tag{103}$$

8.3 MAGNETIZATION

The *magnetization M* is defined as the *magnetic moment per unit volume*. In SI system, the unit of magnetization is given by:

$$1\frac{A.m^2}{m^3} = 1\frac{A}{m} \tag{104}$$

This equation indicates that the unit of magnetization is numerically and physically equivalent to that of the magnetic field, which is consistent with the *Sommerfeld convention* in SI system given by Eq. (75):

$$B = \mu_0(H + M) \quad (SI) \tag{75}$$

In this last equation $\mu_0 = 4\pi \times 10^{-7}$ H/m is the permeability of free space, B is given in *tesla* (T), and H and M in A/m. In *cgs* units, the magnetization is expressed in *emu/cm³*. The relation between the *SI* and *cgs* units of magnetization is derived from Eq. (103), where we obtain the relation:

$$1\frac{emu}{cm^3} = \frac{10^{-3}A.m^2}{10^{-6}m^3} = 10^3\frac{A}{m} = 1\frac{kA}{m} \tag{105}$$

The *specific magnetization σ*, on the other hand, is defined as the *magnetization per unit mass*. In *SI* system, the specific magnetization is expressed in $A.m^2/kg$. We therefore obtain the conversion relation:

$$1\frac{A.m^2}{kg} = \frac{10^3emu}{10^3g} = 1\frac{emu}{g} \tag{106}$$

8.4 MAGNETIC FLUX DENSITY (B)

Maxwell's equation related to Faraday's law is given in *SI* system by:

$$\nabla \times \vec{E} = -\frac{\partial \vec{B}}{\partial t} \quad (SI) \tag{107}$$

Accordingly, an *electromotive* force (*emf*), in units of *volts*, develops in a loop as a consequence of the time-varying flux density (in units of *Tesla*) normal to the area of the loop. This is given by:

$$\varepsilon = -\frac{d\Phi}{dt} \qquad (volts) \tag{108}$$

Here the electromotive force (in *volts*) is expressed as the negative of the rate of change of the magnetic flux:

$$\Phi = \int \boldsymbol{B} \cdot \boldsymbol{da} \tag{109}$$

The *SI* unit of magnetic flux (the number of lines of force) is the *weber* ($1\ Wb = 1\ T.m^2 = 1\ V.s$). In *Gaussian* units, however, Maxwell's equation is:

$$\vec{\nabla} \times \vec{E} = -\frac{1}{c}\frac{\partial \vec{B}}{\partial t} \qquad (Gaussian) \tag{110}$$

Integrating this equation over the area bounded by a closed loop gives the *emf* developed around the loop in units of *statvolt*:

$$\varepsilon = -\frac{1}{c}\frac{d\Phi}{dt} \qquad (statvolts) \tag{111}$$

Here the rate of change of flux is expressed in units of *Mx/s*. The relation between the units of *emf* in the *Gaussian* system (*statvolt*) and the *SI* system (*volt*) is derived from the form of the potential due to a point charge in these two systems.

$$\varphi = \frac{q}{r} \qquad (Gaussian) \tag{112}$$

$$\varphi = \frac{q}{4\pi\varepsilon_0 r} = 9 \times 10^9 \frac{q}{r} \qquad (SI) \tag{113}$$

Accordingly, a point charge of 1 esu produces a potential of 1 statvolt at a point a distance 1 cm away from the point charge. This potential (in volts) is given by Eq. (113) and Eq. (93):

$$1\ statvolt = \frac{9 \times 10^9 \times \left(\frac{1}{3 \times 10^9}C\right)}{10^{-2}m}\ volts = 300\ volts \tag{114}$$

Accordingly, Eq. (111) gives the emf in volts as:

$$\varepsilon = -10^{-8}\frac{d\Phi}{dt} \qquad (volts) \tag{115}$$

We should be reminded here the rate of change of the flux is expressed in Mx/s.

Comparing Eq. (108) with Eq. (115) we realize that the same emf of 1 volt develops from a flux rate of change of 1 Wb/s or 10^8 Mx/s. This leads to the conversion relation for the magnetic flux in the two systems of units (the *SI* and the *emu* systems).

$$1\ Wb = 10^8\ Mx \tag{116}$$

The *magnetic flux density B* is given in units of *Gauss* (*G*) in the cgs system. Accordingly, using the relation in Eq. (116) leads to the conversion relation for the *magnetic flux density,* where:

$$1\ T = 1\frac{Wb}{m^2} = \frac{10^8\ Mx}{10^4\ cm^2} = 10^4\ G \tag{117}$$

8.4 ENERGY PRODUCT

The unit of the energy product (BH) is defined in any system of units by the unit of magnetic induction multiplied by the unit of the magnetic field. In the cgs system, therefore, the unit of the energy product is GOe. Using the conversion relations in Eq. (102) and Eq. (117) we realize that 1GOe = 10^{-4} (T).($10^3/4\pi$)(A/m) =($1/40\pi$) (J/m^3). The commonly used unit of the energy product in cgs system is MGOe = 10^6 GOe. Accordingly, we have the conversion relation:

$$1\ MGOe = 10^6 \times \frac{1}{40\pi}\frac{J}{m^3} = \frac{100}{4\pi}\frac{kJ}{m^3} = 7.96\frac{kJ}{m^3} \tag{118}$$

REFERENCES

[1] A.H. Morrish, The Physical Principles of Magnetism, John Wiley& Sons, New York, 1965.

[2] K.H.J. Buschow, F.R. Boer, Physics of magnetism and magnetic materials, Springer, 2003.
http://dx.doi.org/10.1007/b100503

[3] W. Nolting, A. Ramakanth, Quantum theory of magnetism, Springer Science & Business Media, 2009.
http://dx.doi.org/10.1007/978-3-540-85416-6

[4] N. Ashcroft, N. Mermin, Solid State Physics WB Saunders Co, Philadelphia, PA, 1976.

[5] S. Chikazumi, Physics of Ferromagnetism 2e, Oxford University Press, 2009.

[6] R.C. O'handley, Modern magnetic materials, Wiley, 2000.

[7]   A. Aharoni, Introduction to the Theory of Ferromagnetism, Clarendon Press, 2000.

[8]   B.D. Cullity, C.D. Graham, Introduction to magnetic materials, John Wiley & Sons, 2011.

[9]   W.E. Henry, Spin Paramagnetism of Cr^{+++}, Fe^{+++}, and Gd^{+++} at Liquid Helium Temperatures and in Strong Magnetic Fields, Physical Review, 88 (1952) 559. http://dx.doi.org/10.1103/PhysRev.88.559

[10]   J.S. Smart, Effective field theories of magnetism, Saunders, 1966.

[11]   S. Elliott, The physics and chemistry of solids, Wiley, 1998.

[12]   P. Weiss, Hypothesis of the molecular field and ferromagnetic properties, J. phys, 6 (1907) 661-690.

[13]   L. Neel, Magnetic properties of ferrites: ferrimagnetism and antiferromagnetism, Ann. Phys, 3 (1948) 137-198.

[14]   D.H. Martin, Magnetism in solids, MIT Press Cambridge, MA, 1967.

[15]   R.M. Bozorth, Ferromagnetism, Wiley-VCH1993.

[16]   M. Kleman, Magnetization Processes in Ferromagnets, in: M. Cyrot (Ed.) Magnetism of Metals and Alloys, North Holland, Amstrdam, 1980, pp. 535-601.

[17]   F.N. Bradley, Materials for magnetic functions, Hayden Book Company, New York, 1971.

[18]   K. Baberschke, Anisotropy in magnetism, in: K. Baberschke, M. Donath, W. Nolting (Eds.) Band-Ferromagnetism, Springer, New York, 2001, pp. 27-45. http://dx.doi.org/10.1007/3-540-44610-9_3

[19]   D. Givord, M. Rossignol, D. Taylor, Coercivity mechanisms in hard magnetic materials, Journal de Physique IV Colloque C3, 2 (1992) 95-104.

[20]   D. Craik, R. Tebble, Magnetic domains, Reports on progress in physics, 24 (1961) 116-166. http://dx.doi.org/10.1088/0034-4885/24/1/304

[21]   D. Clerk, Magnetism: Principles and Applications, Wiley, New York, 1995.

[22]   R. Pearson, Magnetic anisotropy, in: G.M. Kalvius, R.S. Tebble (Eds.) Experimental Magnetism, Wiley, Chichester, 1979, pp. 138-223.

[23]   N. Akulov, Zur atomtheorie des ferromagnetismus, Zeitschrift für Physik, 54 (1929) 582-587.

http://dx.doi.org/10.1007/BF01338489

[24] B. Tanner, Antiferromagnetic domains, Contemporary Physics, 20 (1979) 187-210.
http://dx.doi.org/10.1080/00107517908219099

[25] A. Hubert, R. Schäfer, Magnetic domains: the analysis of magnetic microstructure, Springer, Berlin, 1998.

[26] K.H.J. Buschow, New permanent magnet materials, Materials Science Reports, 1 (1986) 1-63.
http://dx.doi.org/10.1016/0920-2307(86)90003-4

[27] E.C. Stoner, E. Wohlfarth, A mechanism of magnetic hysteresis in heterogeneous alloys, Philosophical Transactions of the Royal Society of London A: Mathematical, Physical and Engineering Sciences, 240 (1948) 599-642.
http://dx.doi.org/10.1098/rsta.1948.0007

[28] G. Bertotti, Hysteresis in magnetism: for physicists, materials scientists, and engineers, Academic Press, Sandiego, 1998.

[29] A.M. Alsmadi, I. Bsoul, S.H. Mahmood, G. Alnawashi, K. Prokeš, K. Siemensmeyer, B. Klemke, H. Nakotte, Magnetic study of M-type doped barium hexaferrite nanocrystalline particles, Journal of Applied Physics, 114 (2013) 243910.
http://dx.doi.org/10.1063/1.4858383

[30] M. Awawdeh, I. Bsoul, S.H. Mahmood, Magnetic properties and Mössbauer spectroscopy on Ga, Al, and Cr substituted hexaferrites, Journal of Alloys and Compounds, 585 (2014) 465-473.

[31] S.H. Mahmood, G.H. Dushaq, I. Bsoul, M. Awawdeh, H.K. Juwhari, B.I. Lahlouh, M.A. AlDamen, Magnetic Properties and Hyperfine Interactions in M-Type $BaFe_{12-2x}Mo_xZn_xO_{19}$ Hexaferrites, Journal of Applied Mathematics and Physics, 2 (2014) 77-87.
http://dx.doi.org/10.4236/jamp.2014.25011

[32] S.H. Mahmood, A.N. Aloqaily, Y. Maswadeh, A. Awadallah, I. Bsoul, M. Awawdeh, H.K. Juwhari, Effects of heat treatment on the phase evolution, structural, and magnetic properties of Mo-Zn doped M-type hexaferrites, Solid State Phenomena, 232 (2015) 65-92.
http://dx.doi.org/10.4028/www.scientific.net/SSP.232.65

[33]   A. Alsmadi, I. Bsoul, S. Mahmood, G. Alnawashi, F. Al-Dweri, Y. Maswadeh, U. Welp, Magnetic study of M-type Ru-Ti doped strontium hexaferrite nanocrystalline particles, Journal of Alloys and Compounds, 648 (2015) 419-427. *http://dx.doi.org/10.1016/j.jallcom.2015.06.274*

[34]   S.H. Mahmood, A. Awadallah, Y. Maswadeh, I. Bsoul, Structural and magnetic properties of Cu-V substituted M-type barium hexaferrites, IOP Conference Series: Materials Science and Engineering, IOP Publishing, 2015, pp. 012008. *http://dx.doi.org/10.1088/1757-899x/92/1/012008*

[35]   A. Awadallah, S.H. Mahmood, Y. Maswadeh, I. Bsoul, A. Aloqaily, Structural and magnetic properties of Vanadium Doped M-Type Barium Hexaferrite (BaFe$_{12-x}$V$_x$O$_{19}$), IOP Conference Series: Materials Science and Engineering, IOP Publishing, 2015, pp. 012006. *http://dx.doi.org/10.1088/1757-899x/92/1/012006*

[36]   A. Awadallah, S.H. Mahmood, Y. Maswadeh, I. Bsoul, M. Awawdeh, Q.I. Mohaidat, H. Juwhari, Structural, magnetic, and Mossbauer spectroscopy of Cu substituted M-type hexaferrites, Materials Research Bulletin, 74 (2016) 192-201. *http://dx.doi.org/10.1016/j.materresbull.2015.10.034*

[37]   W.F. Brown, Micromagnetics, Interscience Publishers, 1963.

[38]   H. Zijlstra, Experimental methods in magnetism, North-Holland Amsterdam, 1967.

[39]   P. Compbell, Permanent magnet materials and their applications, Cambridge University Press, Cambridge, 1994. *http://dx.doi.org/10.1017/CBO9780511623073*

CHAPTER 2

High Performance Permanent Magnets

S.H. Mahmood

Physics Department, The University of Jordan, Amman, Jordan

s.mahmood@ju.edu.jo

Abstract

Recent industrial and technological advances led to an exponentially growing demand for high performance permanent magnets. Hard ferrite magnets discovered in the early 1950s were superior to older magnets in terms of magnetic hardness. However, ferrite magnets are relatively weak in terms of flux density and maximum energy product. The need for higher performance magnets subsequently led to the development of rare-earth based magnets with significantly higher coercivity, flux density, and maximum energy products. However, due to the cost effectiveness of ferrite magnets, they dominated the global market of permanent magnet production since the 1980s.

Keywords

Permanent Magnets; Ferrite Magnets; Sm-Co Magnets, Nd-Fe-B Magnets; Coercivity

Contents

1. INTRODUCTION

A magnetic field is needed to perform special tasks in machines and devices. This field can be produced either by an electromagnet (a current passing through a conductor), or by a permanent magnet. Permanent magnets (PMs) have replaced electromagnets in a wide range of applications due to their superiority in constructing machines and devices. They provide higher efficiency and power or torque density, and better dynamic performance of motors than electromagnets. The energy stored in a permanent magnet (by the initial magnetization) is not drained by repeated use, and therefore, an ideal permanent magnet is a perfect power source. In addition, the ease in construction and maintenance, and low cost of production and operation provide additional vital advantages of PMs.

The demand for PMs was driven by the booming industrial and technological advancements. The market demand for PMs increased from 15 k tons (valued at $125 million) in 1964 [1], to 231 k tons (valued at $1.46 billion) in 1987, and to 252 kilo tons (valued at $ 1.755 billion) in 1988 [2]. The global consumption of PMs increased up to 650 kilo tons (valued at $ 15.32 billion) in 2012 according to a recent report published in 2014 [3]. According to the report, the global consumption of PMs is expected to increase up to 1168.7 kilo tons ($28.7 billion) in 2019. An earlier report, however, predicted that the growth in PM global consumption will reach $18.8 billion by 2018 [4]. The results of research on the assessment and estimation of the growth pattern of PM global production (in k tons) and global market sales (in billions of $US) are presented in Fig. 1. The faster increase in sales value compared to the increase in market volume between 1988 and 2012 is largely due to the increased level of production of the more expensive rare-earth based modern magnets.

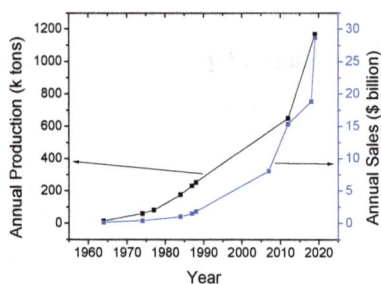

Fig. 1: PM production and sales, with predicted figures in the near future.

The leading sector responsible for the growing demand for PMs is the automotive industry, which accounted for 50% of the volume of the PM market in the year 2012. The electronics sector, the second leading sector, accounted for 25% of the market volume of PM in that year. Further, the growing energy generation industries, specifically, the growth of the wind power industry, is expected to increase the demand for PMs up to 90.5 kilo tons in the year 2019.

Each car uses over 1 kg of PMs, and with millions of cars produced each year, the cost reduction of PMs becomes a critical factor in the PM market. During the last couple of decades, the Chinese government had adopted strategies to cut down the cost of production of PMs. The development of a full scale competitive advantage in mining for rare-earths, and production and processing technologies of PMs, combined with the human capital and low wages, had recently positioned the Chinese at the top of the list of PM producers worldwide. In 1988, the Chinese contribution to the PM market was insignificant compared to Japan and the western countries as illustrated by Fig. 2. However, the Chinese manufacturers became the leading PM producers in 2012, accounting for 62% of the market volume, followed by Asia Pacific countries (including India, Japan, South Korea, and Indonesia) with 15% of the market volume, and then North America with 6% of the market.

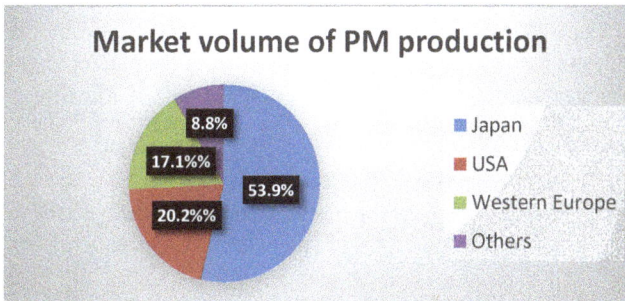

Fig. 2: Regional distribution of PM production in 1988 [2].

The transformations in the global PM market were driven by the Chinese strategies concerning the low-cost production of rare-earth elements essential for modern high-performance magnets, as well as other key factors which enabled Chinese producers to provide low-cost PMs. For example, in 2003, the price of NdFeB magnets was 36

USD/kg in China, and 80 – 82 USD/kg in Japan, USA, and Europe. This could be the key factor in increasing the Chinese production of this type of magnet from 68.6% of the global market in 2003 up to 78.5% in 2008, and the continuous decline of the market shares of Japan, USA, and Europe as illustrated by Fig. 3 [5]. By 2011, it was reported that RE magnets were selling by Chinese producers for under $16 per kg, which could not be matched by producers in other regions, driving the Chinese to have 90% of the market share of RE magnets [6].

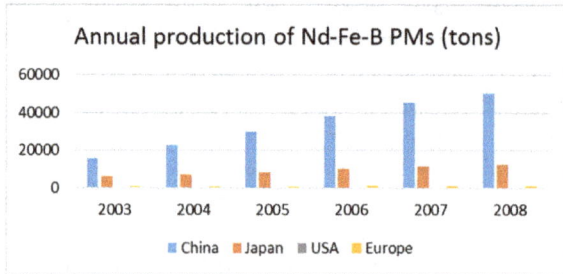

Annual production of Nd-Fe-B PMs (tons)

Fig. 3: Regional distribution of Nd-Fe-B PM production between 2003 and 2008

2. BASICS OF PERMANENT MAGNET OPERATION

Permanent magnets operate on open circuits to provide a magnetic field in a region of space. The free poles induce a *demagnetizing field* (H_d) inside the magnet, opposite and proportional to the magnetization producing it as illustrated by Fig. 4. The demagnetizing field is given by:

$$H_d = -N_d M \tag{1}$$

Here N_d is the *demagnetizing coefficient*, which depends on the magnet geometry.

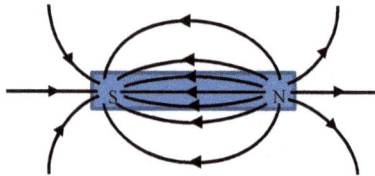

Fig. 4: The H-field lines of a bar magnet.

In cgs units, the magnetic flux density (induction) inside the magnet in the absence of an external applied field is given by:

$$B = H_d + 4\pi M \quad \text{(cgs)} \tag{2}$$

This equation indicates that H_d lowers the flux density in the magnet, and results in a divergence of the flux lines toward the ends of the magnet due to the fact that the H field is stronger near the poles. Unlike the H-field lines, the lines of magnetic flux (B-lines) are continuous as Fig. 5 shows, and outside the magnet, they are identical to the H-lines.

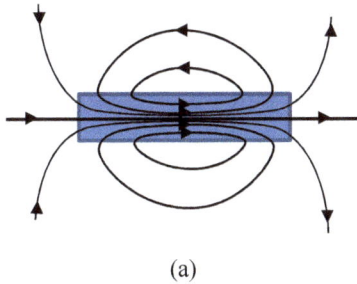

(a)

Fig. 5: The B-field lines of a bar magnet.

Equations (1) and (2) can be combined, resulting in the following equation for the *load line*:

$$B = -\frac{(4\pi - N_d)}{N_d} H_d \quad \text{(cgs)} \tag{3}$$

The intersection between the load line (OL) and the demagnetization curve (the second quadrant in the hysteresis loop) determines the operating point P of the magnet as illustrated by Fig. 6. Notice that the slope of the load line can be changed by the magnet designer, allowing the possibility of fabricating a magnet with the desired operating point.

The magnetic energy density stored in the field produced by the magnet in the space outside the magnet is proportional to $B_d H_d$, the *energy product* of the magnet. This is the figure of merit for the performance of the magnet. Therefore, for practical purposes, it is desired to construct a magnet with geometry that maximizes its BH value. Fig. 7 is a schematic diagram of the energy product of a bar magnet, side by side with its demagnetization curve. The figure demonstrates that the magnet geometry is not optimal, and the operating point can be moved to the point corresponding to the maximum energy product by constructing a longer magnet with the same cross section, or a thinner magnet with the same length. This is due to fact that increasing the axial ratio reduces N_d, and moves the operating point upward toward the point of maximum energy product (see Eq. (3).

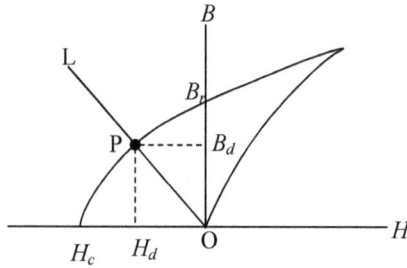

Fig. 6: The magnetization and demagnetization curves in the first and second quadrants of the hysteresis loop, and the load line of a permanent magnet; B_d and H_d define the induction and demagnetization fields at the operating point of the magnet.

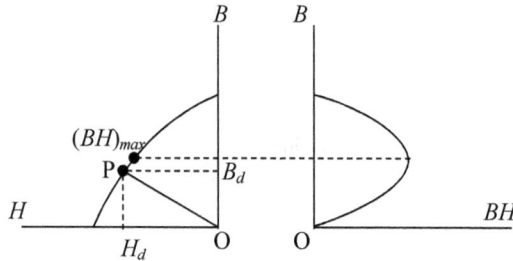

Fig. 7: Demagnetization curve (left) and the corresponding values of the energy product (right).

The coercivity force, which is the reverse field required to reduce the induction or the magnetization to zero, is also an important index of the quality of the permanent magnet. The reverse field required to remove the induction is termed *coercivity* (H_c, H_{cB}, $_BH_c$), while that required to remove the magnetization is called *intrinsic coercivity* (H_{cJ} or H_{ci}). The intrinsic coercivity of a PM material is proportional to its magnetocrystalline anisotropy. Therefore, materials with high magnetocrystalline anisotropy are suitable for high performance PMs.

Equation (2) indicates that when the H-field is small relative to $4\pi M$ as in the case of a soft magnetic material, the B vs H curve becomes almost identical to the plot of $4\pi M$ vs H. For a hard magnetic material where M remains constant far into the second quadrant (material with high intrinsic coercivity), such as in the case of aligned hard ferrite magnet, the B-H curve becomes significantly different from the $4\pi M$-H curve as demonstrated by Fig. 8. Specifically, the coercive field H_c (also recognized as H_{cB} or $_BH_c$) is significantly lower than the intrinsic coercivity H_{cJ} (or H_{ci}). Notice that the knee in the B-H demagnetization curve is generally a disadvantage, since it reduces the coercivity of the magnet material, and lowers its performance at high temperatures. In this respect, the advantage of hard ferrite material over other magnet materials, however, is that its intrinsic coercivity increases at elevated temperatures.

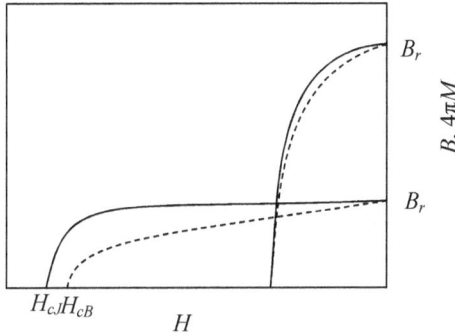

Fig. 8: Schematic diagram of the demagnetization curves of a ferrite magnet and Alnico magnet. The solid lines represent the $4\pi M$-curves, while the dashed lines represent the B-curves.

It is noteworthy mentioning that the squareness ratio (the remanent magnetization divided by the saturation magnetization, M_r/M_s) of an isotropic magnet is ~ 0.5. The remanence can be significantly improved by the construction of anisotropic (aligned) magnet of the material. In an ideal perfectly aligned magnet, the squareness ratio is 1, which leads to significant improvement of the magnetic properties of the magnet in terms of the residual flux remanence B_r and energy product.

A permanent magnet with high magnetic properties is characterized by a linear B-H curve in the second quadrant. This can be achieved by fabricating a perfectly aligned magnet from a material with very high magnetocrystalline anisotropy, and (ideally) rectangular B-H demagnetization curve as illustrated by the schematic diagram in Fig. 9.

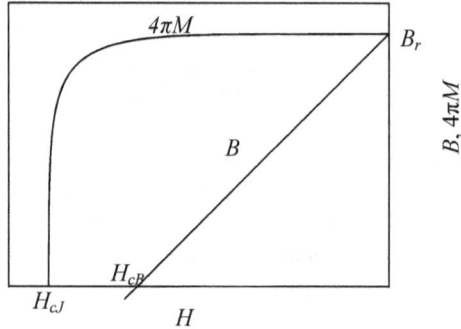

Fig. 9: Schematic diagram of the demagnetization curves of an ideal hard magnet.

Under these circumstances, the remanence magnetization is equal to the saturation magnetization, which is the highest achievable remanence. Also, the coercivity of such a magnet would be maximum:

$$H_{cB} = B_r = 4\pi M_s \tag{4}$$

Accordingly, the maximum energy product, which is the maximum area inscribed by a rectangle in the second quadrant is given by:

$$(BH)_{MAX} = \frac{B_r}{2} \cdot \frac{H_{cB}}{2} = \frac{B_r^2}{4} \tag{5}$$

With B_r given by equation (4), this value would be the highest (theoretical) achievable maximum energy product. For example, the theoretical maximum energy product of a densified barium hexaferrite magnet with optimal saturation magnetization of 72 emu/g would be 5.8 MGOe (46 kJ/m^3).

Of course there is no benefit in having $H_{cJ} \gg H_{cB}$, as far as the magnetization curve remains almost constant far enough into the second quadrant to insure linearity of the B-H curve. In fact, having a very high H_{cJ} could be a disadvantage, since the magnet would require a very high external field for charging (initial magnetization). This is commercially translated in terms of raising production cost. Therefore, tuning the magnetic properties of the magnet material by balancing the reduction of the intrinsic coercivity and the enhancement of the residual flux density is crucial for high performance magnet design.

3. DEVELOPMENT OF COMMERCIAL MAGNETS

The oldest known material for permanent magnets is the *loadstone* used in ancient compasses invented in China more than 3000 years ago. This material, known for about 5000 years, is mainly composed of Fe_3O_4 (magnetite) mineral. Magnetite is magnetically soft that can be demagnetized by a relatively small external reversing field. Accordingly, this material cannot be used as efficient permanent magnet. The need for magnets with better mechanical and magnetic properties instigated serious research vigorous devoted to the development of high performance permanent magnets. These efforts led to the discovery of the high performance permanent magnets of today. The most important commercial permanent magnets are discussed below, together with the first magnet (carbon steel) produced for practical applications.

3.1 CARBON STEEL MAGNETS

The production of the quench hardened iron-carbon alloys (sword steel) was the fruit of the experimental work of William Gilbert (1540 – 1603) on magnetism, which appeared in his book *On the Magnet*, published in 1600. Attempts to develop iron magnets for more than 150 years after Gilbert's work led to improvements in the production of PMs for applications, but with no important discoveries in the field. In the eighteenth century, steel magnets made of magnetized steel strips were used to lift iron parts. In that time, magnetizing the strips was done by rubbing with loadstone. However, the onset of real research on magnetic materials had to await the 1820 great discover of Hans Christian Oersted (1775 – 1851) that a magnetic field can be produced by an electric current, and the subsequent development of the first electromagnet in 1825 [7].

Carbon steel magnets have saturation magnetization superior to magnetite. However, the low coercivity of carbon steel, where a weak reversing field of only 50 Oersted is able to demagnetize the material, is an obvious disadvantage for operation in the presence of stray fields. The low energy product of < 0.25 MGOe (2 kJ/m^3) is another disadvantage. After 1880, alloying the steel with other metals was found efficient in improving the magnetic properties of the steel, and by 1885, steel added with 5% tungsten was produced and used for magnets. This was replaced by the cheaper chromium steel during World War I. However, until 1900, the coercivity of tungsten and chromium steels was still below 100 Oe, and their energy product did not exceed 0.3 MGOe (2.4 kJ/m^3).

Continuous efforts to improve the magnetic properties of steel magnets led to the discovery that alloying with cobalt could triple the coercivity. In 1917, Kotaro Honda filed three patents on the invention of new carbon steels with enhanced magnetic properties, and the invention of carbon steel with 35% Co was patented in 1920 [8]. Carbon steels with 30 – 40% cobalt, in addition to tungsten and chromium were reported with enhanced energy product up to 1 MGOe (8 kJ/m^3), and with the highest recorded coercivity of a magnetic material (230 Oe). However, the magnetic properties of steel magnets were far inferior to those developed a decade later. In addition, the best steel magnets are costly due to their high content of the expensive cobalt element. Consequently, the production of these magnets on a commercial scale was discontinued and replaced by modern, high performance magnets.

3.2 ALNICO MAGNETS

The patent of the first magnet alloy based on Fe, Ni and Al (Alni) by Tokuhichi Mishima in 1931 signaled the shift of the attention of the scientific community toward the development and production of a new type of materials for high performance magnets. Mishima reported that the new $Al_{12}Ni_{30}Fe_{58}$ alloy had a coercivity of 400 Oe, almost double that of the best steel magnet. The magnetic properties of Mishima magnet were improved by the introduction of new elements, mainly Co, and small amounts of Cu and possibly other elements. The 3-decade period of development of these materials culminated in the production of magnets with coercive forces $H_{cB} \sim 1900$ Oe (150 kA/m) and energy products exceeding 9.5 MGOe (76 kJ/m^3), thus boosting the use of permanent magnets in the machine industry. In addition, alnico magnets have the advantages of high remanence (maximum $B_r = 1.33$ T), and high Curie temperature ($T_c = 850°$ C) which makes them usable in high-temperature regimes ($\geq 500°$ C).

Alnico alloys are hard and brittle, and cannot therefore be cold-worked. Consequently, the production processes of alnico magnets are limited to either melting and casting, or grinding to fine powder, pressing, and sintering. The final finishing process is limited to surface grinding of the magnet. Sintered magnets have finer grains, and are mechanically stronger and have better surface finishing than the cast magnets. However, the magnetic properties of sintered magnets are somewhat inferior to cast magnets. The as-cast or as-sintered magnets have relatively poor magnetic properties, which can be improved by special heat and magnetic field treatments to homogenize the solid solution and increase the remanence and coercivity.

The major advantage of alnico magnets is their high remanence and high-temperature performance. Alnico products with an operating temperature of 500° C are now commercially produced with the following magnetic properties: $B_r = 1.12$ T, $H_{cB} = 150$

kA/m, and $(BH)_{max} = 92$ kJ/m^3. However, the requirement of higher coercivity and energy products for device miniaturization, in addition to the poor mechanical properties (brittleness and hardness), and relatively high market price of cobalt had limited the growth of production of such magnets.

3.3   Sm-Co MAGNETS

Magnets with significantly improved magnetic properties were realized by the introduction of the rare earth-transition metal (RE-TM) alloys series. The 1966 discovery of the high magnetocrystalline anisotropy in YCo$_5$ instigated research devoted to the development of Re-Co intermetallic compounds for PM applications [9]. SmCo$_5$ was discovered in mid 1960s by Karl Strnat with energy product 143 kJ/m^3, far superior to other existing permanent magnets. The product was introduced as the first generation of high performance (1–5) RE magnets in 1970. This compound was found to have a significantly higher coercivity than YCo$_5$ and other RE-Co derivatives [9, 10]. Although the Curie temperature of this intermetallic is $T_C \sim 750°$ C, the behavior of the coercivity with increasing temperature limits their use to below 250° C. Such working temperature are reasonably high, but inferior to Alnico magnets. Also, the high cost of this magnet due to the high content of the expensive Sm metal (34 wt. %) is a disadvantage.

The properties of the 1–5 magnets were modifies by adopting different processing technologies, and RE and TM substitution scenarios. Partial substitution of Sm by Pr results in a slight reduction of the cost, and improvement of the energy product, but this substitution reduces the coercivity of the product. In early 1970s, and only few years after the production of sintered SmCo$_5$ with energy product of just over 160 kJ/m^3, Sm-Pr-Co$_5$ magnet with energy product of about 200 kJ/m^3 was produced. A significantly higher energy product of 239 kJ/m^3 was achieved by additions of limited amounts of Cu, Zr, and Fe [11]. Replacing Sm by Ce cuts down the cost, but it reduces the energy product and the curie temperature of the product. Complex mixtures of Sm(Ce,Pr)-Co(Fe,Cu) "1-5" magnets were reported with B$_r$ = 0.50 – 1.03 T, H$_{cj}$ = 398 – 1990 kA/m, and $(BH)_{max}$ = 47.8 – 207 kJ/m^3 [2]. Better performance at higher temperatures was achieved by the introduction of heavy rare earth elements (such as Gd), but this increased the magnet cost, and lead to reduction of the remanence, and lowering the energy product down to the range 72 – 111 kJ/m^3. The magnet cost can also be reduced by Cu substitution for Co, which allows for the introduction of Fe into the product, but this process has the disadvantage of lowering the remanence and energy product. 1–5 magnet with optimum composition of 31 wt. % Sm, 46 wt. % Co, and 23 wt. % Cu was reported to have: B$_r$ = 0.61 T, H$_{cB}$ = 279 kA/m, and $(BH)_{max}$ = 55 kJ/m^3 [12].

Modification of the properties of the 1–5 magnets can therefore be achieved by several choices of metal substitutions for Sm and Co, but it seems difficult to achieve advantages in all critical parameters and cost of the product. A variety of 1–5 magnets with upper working temperatures of 200 – 350° C are now commercially produced with the following magnetic properties: B_r = 0.59 – 1.0 T, H_{cJ} = 358 – 2000 kA/m, H_{cB} = 358 – 770 kA/m, and $(BH)_{max}$ = 68 – 190 kJ/m^3.

A distinct advantage of SmCo$_5$ magnets is the ability to be charged (magnetized) in a field much smaller than their intrinsic coercivity, cutting down the production cost. This is due to the magnetization-demagnetization and the nucleation-controlled coercivity behavior of the magnet as illustrated by Fig. 10. The initial magnetization is accomplished by domain wall motion, which requires relatively small applied fields. The demagnetization curve is almost flat until a reverse field close to the intrinsic coercivity is applied, where the magnetization drops abruptly. Magnetization reversal in SmCo$_5$ occurs by nucleation of reverse domains which requires a high field. The nucleation field is high enough to move the domain walls in the grains and attain saturation in the opposite direction [7].

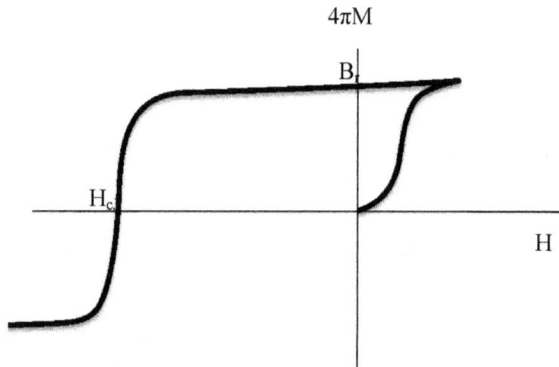

Fig. 10: Schematic diagram illustration of the magnetization-demagnetization curve of SmCo$_5$.

The second generation of RE magnets is the 2–17 family introduced to the market in 1980 with the nominal commercial composition Sm(Co, Fe, Cu, Zr)$_{7.4}$. This magnet has a lower Sm content than the 1-5 family, and is therefore expected to cost less. However, the higher processing cost balances the saving in raw material costs, and keeps the market price of this magnet at the higher end of the price list of PMs. Although the brand name "2-17" is used for these magnets, the commercial products of this family is believed to be composed of regions of the 2-17 phase separated by bands of the 1-5 phase [7]. The mechanism of producing high coercivity in this magnet is different from that in $SmCo_5$; it results from domain wall pinning rather than from domain nucleation as in the case of $SmCo_5$. The 2–17 magnet family has in general higher B_r than the 1–5 family, and therefore higher energy product ($159 - 239$ kJ/m^3) [2].

Continuous research efforts led to the production of of a variety of 2-17 magnets with significantly improved coercivity, remanence, and energy products. In 1990, Liu & Ray reported the production of a magnet containing light RE metal, with nominal composition $Sm_{0.8}(Ce_{0.2}Pr_{0.4}Nd_{0.4})_{0.2}(Co_{0.633}Fe_{0.286}Cu_{0.061}Zr_{0.020})_{7.59}$; the magnetic properties of the product: $B_r = 1.157$ T, $H_{cJ} = 1234$ kA/m, $H_{cB} = 822$ kA/m, and $(BH)_{max} = 239$ kJ/m^3 [13]. However, the highest energy product of 271 kJ/m^3 was achieved by using higher Fe content in Sm-Co compound with nominal composition $Sm(Co_{0.612}Fe_{0.316}Cu_{0.052}Zr_{0.020})_{7.73}$ [14]. Further, the intrinsic coercivity of the compound $Sm(Co_{bal}Fe_{0.06}Cu_xZr_{0.03})_z$ was found to be sensitive to x and z values, being more sensitive to z [15]. It was reported that for $x = 0.1$ and 0.12, the intrinsic coercivity at $z = 5.8$ was 438 kA/m, and increased dramatically to 3184 kA/m at $z = 7.2$. However, at a lower Cu content of $x = 0.8$, the intrinsic coercivity increased from 119 kA/m at $z = 5.8$ up to 1880 kA/m at $z = 7.2$. Sm_2Co_{17} magnet with high magnetic properties and upper working temperatures in the range $200 - 350°$ C were commercially produced. Different commercial grades of 2–17 magnets are now available with magnetic parameters $B_r = 0.93 - 1.15$ T, $H_{cJ} = 636 - 1990$ kA/m, $H_{cB} = 557 - 845$ kA/m, and $(BH)_{max} = 160 - 255$ kJ/m^3.

The use of Sm-Co magnets for high-temperature applications is limited by the deterioration of their performance above $300°$ C. Serious efforts were then made to produce Sm-Co magnets with higher working temperatures. The intrinsic coercivity of $Sm(Co, Fe, Cu, Zr)_z$ ($z = 6.7 - 9.1$) was investigated at high temperatures [16]. In that work, Liu et al. produced magnets with the highest ever reported coercivity of more than 796 kA/m at $500°$ C. Further, laboratory success was made in producing $Sm(Co_{6.1}Cu_{0.6}Ti_{0.3})$ with coercivity of 685 kA/m at $500°$ C [17]. This compound was reported to be a two-phase mixture of 1–5 and 2–17 Sm-Co phases. The residual flux of this high-temperature magnet, however, was inferior to Alnico, and its energy product (31.8 kJ/m^3) was significantly lower than that for other commercially produced Sm-Co magnets. The high-temperature magnetic properties were marginally enhanced by the introduction of Fe, where the intrinsic coercivity at $500°$ C was reported to be 764 kA/m for $Sm(Co_{6.0}Fe_{0.4}Cu_{0.6}Ti_{0.3})$, and improvement of the energy product and magnetization was achieved [18].

Commercial $SmCo_5$ permanent magnets are manufactured by powder-metallurgical technique from alloys synthesized by one of the following methods [2]:

(1)   Induction melting and casting the metallic constituents in a crucible.

(2)   Calcio-thermic reduction-diffusion (R-D).

In the first process, the as-cast alloy is crushed and milled to produce micron-size particles. The particles are aligned in a magnetic field (for anisotropic magnets), and pressed in a die. The compacts are then sintered at temperature $\gtrsim 1000°$ C. A small amount of powder prepared with excess amounts of Sm is sometimes added to the magnet powder before compaction. This powder melts at the sintering temperature, and the *liquid-phase sintering* process results in higher density and better magnetic properties. The mean grain size of the sintered alloy is an order of magnitude larger than the critical single-domain particle size, and multi-domain structure in the alloy makes it possible to charge the magnet with a field significantly smaller than the magnet coercivity as mentioned previously [7]. The PM production process is completed by cutting or grinding into the final desired shape, and re-magnetizing if necessary.

In the second process (also known as KOR process), the RE, and possibly some of the transition metals are used in oxide powder forms, while the Co and Fe are introduced as metal powders. The calcio-thermic reduction of oxides with calcium results in a spongy alloy powder by simultaneous diffusion of the powder constituents. This process lowers the production cost of RE magnets by eliminating several steps. The powders are then treated as in the first process to achieve final products. The high chemical reactivity of the RE and alloy powders requires great care and the use of vacuum or an inert atmosphere during processing.

The magnetic powders could also be used to manufacture bonded magnets, where appropriate mixtures of the magnetic material with a binder (a plastic, rubber, or a low-melting point material) are molded or pressed. Subsequent heat treatment is used for the curing step. Finally, magnets with desired shapes are easily made due to the ductility of the material. The production of RE-bonded magnets dates back to the mid-1970s, but did not receive much attention due to their weaker magnetic properties compared to sintered magnets. Later in the 1980s, the Japanese efforts succeeded in placing boded RE-magnets in the list of popular magnets. Early SmCo5 bonded magnets were produced with energy products of 3–10 MGOe ($24 - 80$ kJ/m^3), which are comparable or higher than the energy products of the best alnico or ferrite magnets produced nowadays.

The manufacture of the 2–17 magnets is essentially similar to the procedure used for manufacturing $SmCo_5$ magnets, except that special heat treatment is required to obtain the microstructure necessary for optimal magnetic properties. The reduction-diffusion process was successfully modified for the production of the 2–17 magnets.

3.4   Nd-Fe-B MAGNETS

Due to the scarcity of Sm, and the high market price of Sm and Co metals, scientist devoted serious efforts in search of cheaper materials with enhanced magnetic properties. In a series of publications early 1980s, John Croat's group at General Motors reported the production of RE-Fe binary alloys with promising magnetic properties for PM applications [19-21]. $Nd_{0.4}Fe_{0.6}$ with intrinsic coercivity of 593 kA/m alloy was reported to have the highest coercivity among Re-Fe alloys [19]. At the same time, Koon and collaborators published the results of their work on $(Fe,B)_{0.9}Tb_{0.05}La_{0.05}$ alloys [22, 23], and reported an intrinsic coercivity of 716 kA/m and a remanence of 0.5 T for the alloy $(Fe_{0.82}B_{0.18})_{0.9}Tb_{0.05}La_{0.05}$. The results of these experiments laid the foundations for the development of the new Nd-Fe-B magnets. The work of Hadjipanayis (at Kollmorgen Company in Radford, Virginia at the time) and collaborators in April 1983 accelerated the pace of research work devoted to the development of cheaper, and more efficient magnets. One of their interesting products was Pr-Fe-B-Si magnetic material prepared by melt spinning and annealing, and an article on the experimental results was submitted to Applied Physics Letters on June 23rd, 1983, and accepted for publication on August 1st of that year. The compound $Fe_{76}Pr_{16}B_5Si_3$ was reported to have a maximum energy product of 95.5 kJ/m^3; an outstanding Co-free PM [24].

In June 1983, the Sumitomo Special Metals Company of Osaka, Japan announced the production of a new magnet by powder metallurgical method with record energy product. In a subsequent publication by Sagawa and collaborators, the structural and magnetic properties of $Nd_xB_yFe_{100-x-y}$ intermetallics (with $x = 13 - 19$, $y = 4 - 17$) were described [25]. These intermetallics contained slightly more RE (Nd) content than Koon's compound $((Tb,La)_{10}Fe_{74}B_{16})$ [22], and significantly less Nd content than Croat's Nd-Fe binaries [19, 20]. In these new compounds, the intrinsic coercivity of $Nd_xB_8Fe_{92-x}$ was found to improve up to $x = 15$, and then slightly improved at higher Nd content, while B_r and BH_{max} were optimal in the range $x = 14 - 15$. On the other hand, the intrinsic coercivity of $Nd_{15}B_yFe_{85-y}$ improved up to $y = 8$, and only slightly improved at greater B-content, while Br and BH_{max} revealed optimal values at $y = 6 - 8$. Specifically, $Nd_{15}B_8Fe_{77}$ was found to crystallize into a tetragonal phase with $a = 0.880$ nm and $c = 1.221$ nm, and enhanced magnetic parameters of $B_r = 1.23$ T, $H_{cJ} = 960$ kA/m, $H_{cB} = 880$ kA/m, and a record energy product of $(BH)_{max} = 290$ kJ/m^3 [25]. The intrinsic coercivity of this compound increased with increasing the maximum magnetizing field strength, reaching $H_{cJ} =\sim 850$ kA/m when a magnetizing field of 600 kA/m was used, and then the squareness improved, and H_{cJ} reached the optimal value at a magnetizing field strength of 1200 kA/m.

On the other side of the Pacific, in November 1983, several American groups discussed their results on the new RE-Fe-B magnetic materials at the 29th Annual Conference on Magnetism and Magnetic Materials in Pittsburgh. Koon and Das reported on the production of Pr-Fe-B and Nd-Fe-B magnets, and announced the achievement of maximum energy product of 103 kJ/m^3 for their annealed melt-spun ribbons. Similarly, General Motor's Croat reports the production of Pr-Fe-B and Nd-Fe-B magnets with energy product of 120 kJ/m^3, and announced that the company had a way to double this value. In their publication in Applied Physics letters in 1984, Croat et al. reported on the production of Nd-Fe-B magnet with energy product of 112 kJ/m^3, which they claimed to be the highest reported energy product for a light RE-Fe material [26]. Further, in 1984 the results of Croat's group on a series of Nd-Fe-B magnets appeared in the same volume of Journal of Applied Physics, and just before the article of Sagawa's group. In the latter publication, Croat et al. reported their results on a series of (Nd,RE)-Fe-B compounds prepared by melt spinning under different experimental conditions [27]. The $Nd_{0.135}(Fe_{0.935}B_{0.065})_{0.865}$ composition gave a remanence of about 8 kG (= 0.64 T), a maximum energy product of 14 MGOe (111 kJ/m^3), and an enhanced intrinsic coercivity of about 15 kOe (1194 kA/m) [27]. Obviously, the Japanese product is superior in terms of remanence and energy product, while the American product is superior in terms of magnetic hardness. Analyzing the variations of the magnetic properties with the variations of Nd and B content in the compounds of the Japanese group, it was clear that the small differences in the compositions of the compounds investigated by the Japanese and the American groups could not have induced the observed significant differences in magnetic properties. Accordingly, the observed differences could indicate the sensitivity of the magnetic properties of Nd-Fe-B magnets to the preparation method and the microstructure of the samples [28].

The Curie temperature of Nd-Fe-B magnet is $T_C \sim 310°$ C, which limits its performance at elevated temperatures in comparison with Sm-Co magnets. Nowadays, high flux, and high energy product NdFeB magnets with B_r = 1.45 T, $H_{cJ} \sim$ 1114 kA/m, H_{cB} = 1080 kA/m, and $(BH)_{max}$ = 414 kJ/m^3 are commercially available with a working temperature up to 100° C. Although some of the early developed Nd-Fe-B magnets had high flux density and high energy products, they were not good candidates for motor applications due to the non-linearity of their B-H curves, and their instability at high temperatures. Thus, real efforts were devoted to the improvement of the high-temperature performance of NdFeB magnets. While a knee was observed in room-temperature B-H curve of Nd-Fe-B magnet, partial substitution of Nd by Dy, ($Nd_{10}Dy_4Fe_{80}B_6$) increased the coercivity, and maintained a linear B-H behavior up to 175° C, thus improving the properties of the magnet for high-temperature applications [5]. This substitution, however, results in reduction of the remanent flux density and energy product, and increases the magnet production cost. Commercial grades of high- coercivity NdFeB-based magnets working at higher temperatures (up to 220° C) are now available with the following magnetic parameters: B_r = 1.27 T, H_{cJ} ~ 2662 kA/m, H_{cB} = 971 kA/m, and $(BH)_{max}$ = 310 kJ/m^3. These magnetic properties are superior to those of Sm-Co magnets, at comparable operating temperatures.

It seems that this was the type of PMs the world has been waiting for; a high performance, Co-free magnet, especially with the unexpected fluctuations in Co-price due to the high market demand, and the political instability in the supplying regions. Accordingly, the production of Nd-Fe-B magnets increased exponentially in the period 1997 – 2002, and the exponential rate of increase was intensified in the period 2003 – 2006. The market sales of this strategic PM increased from \$1.142 billion in 2003 up to \$3.436 billion in 2008 [5]. Fig. 11 demonstrates the growth pattern in the global annual production of Nd-Fe-B magnets.

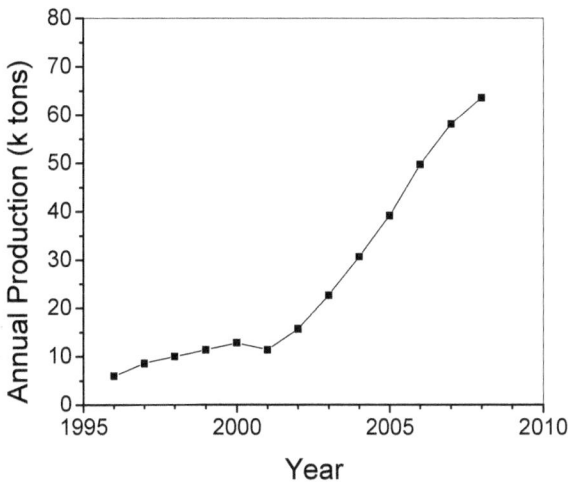

Fig. 11: Global annual production of Nd-Fe-B PM between 1996 and 2008.

In principle, Nd-Fe-B magnets can be produced by the same methods adopted for the production of Sm-Co magnets. Indeed, the original alloy of the Sumitomo Special Metals was produced by the induction melting method. However, due to difficulties associated with the commercial production of NdFe-B by R-D/KOR methods, an alternative production method was developed by the General Motors Corporation. This method is based on the rapid solidification of the molten alloy by melt spinning technique. The melt solidifies as flakes with small crystallites (~ 10 nm diameter), which are subsequently milled into platelet-shaped particles, and annealed to produce a very high-coercivity powder. The resulting isotropic finely polycrystalline powder particles cannot be used for the production of anisotropic magnets by aligning the particles by a magnetic field. Isotropic bonded magnets, however, can be, and are indeed, commercially produced by mixing the magnetic powder with a binder (such as polymer or rubber), molding or pressing, and then curing [2]. Even though the binder dilutes the magnetic properties significantly, most RE-based bonded magnets remain superior to most non-earth magnets; this is a consequence of the high magnetic properties of the RE magnets. Also, densification without magnetic alignment is achieved by hot pressing. In addition, the developments in hot deformation techniques made it possible to improve the magnetic anisotropy, and produce well-oriented magnets with energy product as high as 40 MGOe (318 kJ/m^3), essentially equivalent to the best sintered magnets [2]. These magnets are more corrosion resistant, and may have better long-term performance than sintered magnets.

3.5 FERRITE MAGNETS

Hexaferrites with properties suitable for permanent magnets were discovered by the Philips group in the 1950s. These ceramic oxides are used in a wide range of electronic and electrical applications due to their high resistivity, and low eddy current losses. Also, they possess desirable characteristic as microwave absorbers, and are used for shielding equipment from electromagnetic noise signals, as well as radar absorbers in military applications. A disadvantage of ferrite permanent magnets as compared to SmCo and NdFeB magnets is their low magnetic properties and energy product. The commercially available high performance ferrite magnets are characterized by the following magnetic parameters: B_r = 0.41 T, H_{cJ} ~ 335 kA/m, H_{cB} = 300 kA/m, and $(BH)_{max}$ = 35 kJ/m^3.

Intensive research efforts over the years of the 20th century have led to the development of high performance modern permanent magnets, and significant improvement of the maximum energy product as illustrated by Fig. 12 [29]. In addition, research efforts led to the improvement of other magnetic properties of permanent magnets essential for a wide variety of practical applications. The ranges of the magnetic properties of the leading commercially available PMs dominating the world market today are presented in Fig. 13.

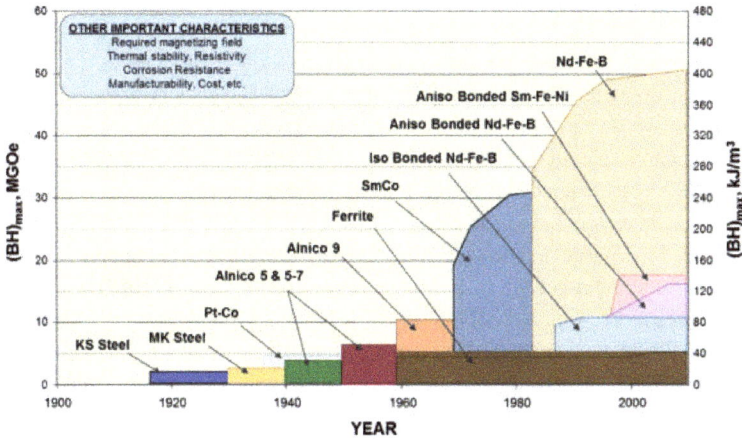

Fig. 12: Historical improvement of the maximum energy product of the various types of permanent magnets produced for practical use [29].

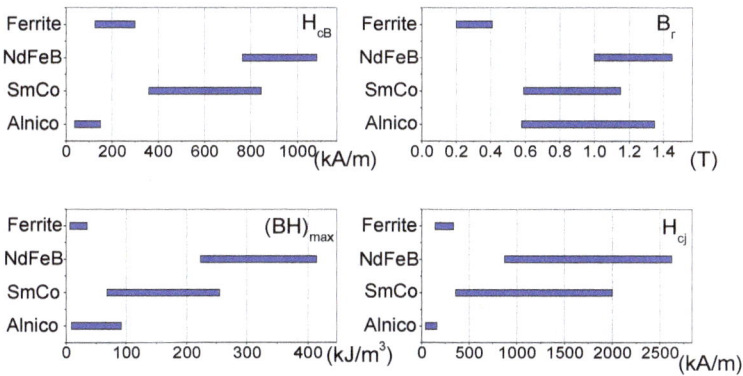

Fig. 13: Magnetic properties of commercially available permanent magnets.

In spite of the disadvantage of inferior magnetic properties compared to RE-based magnets, ferrite magnets have advantages such as low production cost, availability of raw materials, and high corrosion resistance at high working temperatures (> 250° C) compared to Re-Co and NdFeB magnets. The scarcity of RE-elements, and the monopoly imposed on the word market by Chinese producers in the post 1985 era (Fig. 14), in addition to the Chinese policy of limiting RE-element export due to internal demand, all these factors led to an exponential increase in RE market prices to almost a prohibitive level. While ferrite sintered magnets are available at ~ $5 - $20 per kg in US European markets, sintered RE-based magnets are sold at much higher prices reaching ~ $200/kg at the higher end of the market price as illustrated by Fig. 15 [29]. For a flux density of about three times more, the market price of RE magnets makes these magnets less cost-effective than ferrite magnets. Further, tripling the surface area of the weaker ferrite magnet can provide flux density comparable to that provided by RE magnets in an air gap. However, this may entail larger machine volume, a problem which can be solved by novel machine design. Further, ferrite magnets can be used in powder form for real applications, cutting down the cost of machining and processing. These advantages of ferrite magnets over RE-based magnets have led to the emergence of ferrite magnets as the leading PMs in the market since 1987.

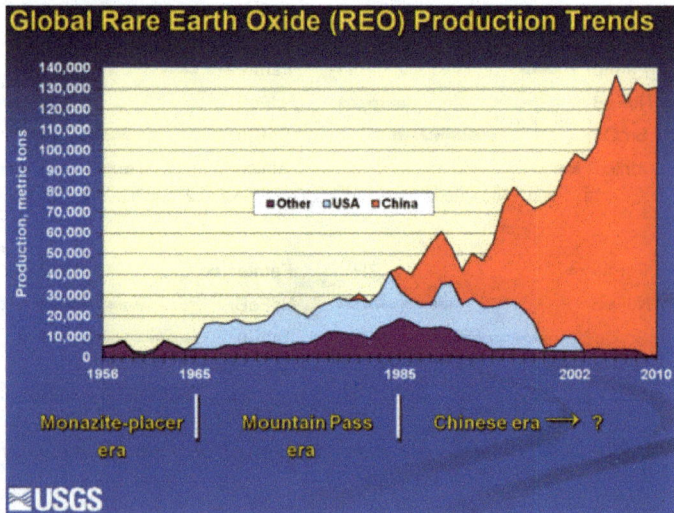

Fig. 14: Illustration of the timeline of production of RE oxides by major producers [29].

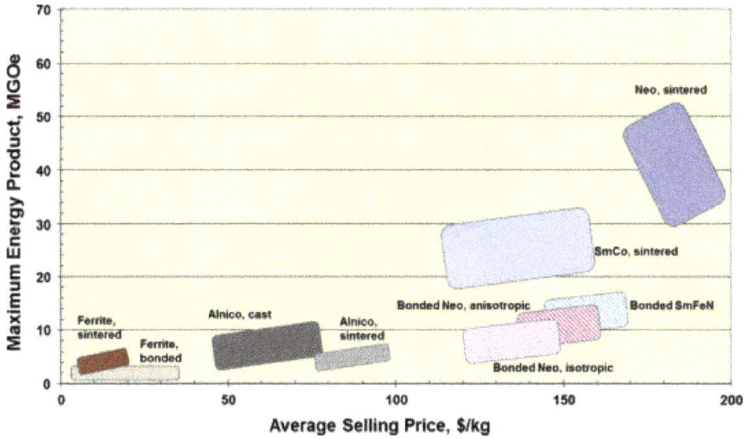

Fig. 15: Maximum energy product versus selling prices of commercially important magnets [29].

According to a recent report published in 2014, ferrite magnets exceeded 80% of the market volume of PMs in 2012 as illustrated by Fig. 16 [30]. Although having higher performance, the high cost of raw materials, and of the production and processing of SmCo and NdFeB magnets seem to be prohibitive. However, due to the low price of ferrite magnets in comparison with RE-based magnets, ferrites with more than 80% of the global PM market volume accounted for only just above 20% of the global PM market revenues, while Alnico and Sm-Co magnets (with only 3% of the global market volume) accounted for about 8% of the global PM market revenues as illustrated by Fig. 17. Obviously, Nd-Fe-B magnets dominated the PM world market in 2012, and the market share of 62% predicted in 2010 have grown significantly in 2012. Despite the high price of Nd-Fe-B magnets, their market share recorded continuous growth since their emergence. This is due to the demand for high performance magnets with high flux and energy product required for device miniaturization.

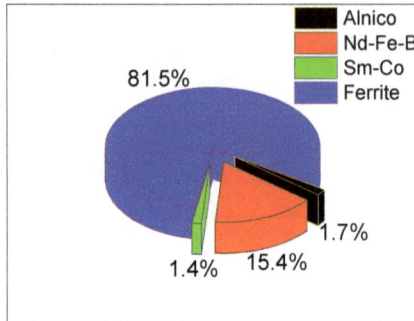

Fig. 16: Market volume share of the most important commercially produced PMs in 2012.

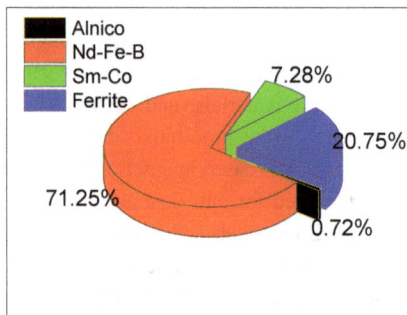

Fig. 17: Market sales of the most important commercially produced PMs in 2012.

The volume of published scientific research and number of patents concerning permanent magnet materials may reflect the degree of interest of scientists and developers, as well as the degree of potential for commercialization of the product. The chart in Fig. 18 illustrates the evolution of published scientific research related to the four most important commercial PMs during the indicated periods. This chart demonstrates exponential growth in published scientific work, consistent with the growing demand for PMs. Of the four types of magnets, Nd-Fe-B magnets dominated scientific research during the last three decades, indicating the great scientific interest in developing and characterizing Nd-Fe-B magnetic materials compared to the other magnetic materials. The chart also shows an exponential growth of the number of published hexaferrite articles since the discovery of the ceramic oxide in the early 1950s.

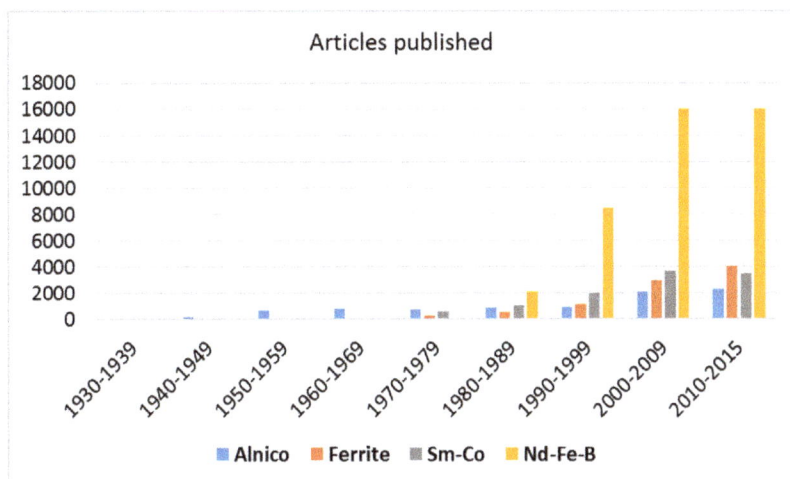

Fig. 18: Number of published articles related to the different commercial PMs.

The chart in Fig. 19 illustrates the number of registered patents during the same periods. It is clear that the number of Nd-Fe-B PM patents grew rapidly since the discovery of these magnets, which is an indicator of the rapidly growing development of these materials during the last three decades. Fig. 20 demonstrates the shares of patents of the four types of commercial PMs. Although discovered fifty years after alnico magnets, Nd-Fe-B magnets recorded similar number of patents now, and is expected to exceed the overall number of patents of alnico in the near future.

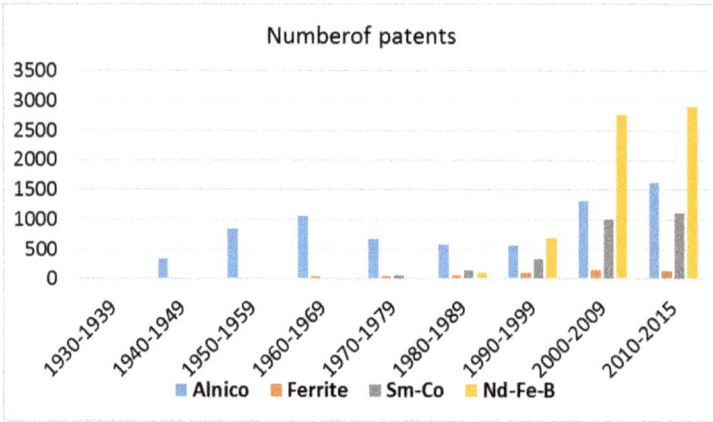

Fig. 19: Number of patents related to the different commercial PMs.

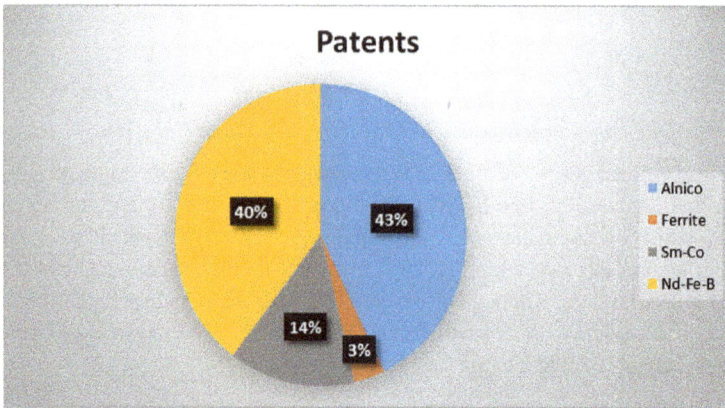

Fig. 20: Percentage share of the number of patents related to the different types of commercial PMs.

REFERENCES

[1] D. Howe, Review of Permanent Magnet Applications and the Potential for High Energy Magnets, Supermagnets, Hard Magnetic Materials, Springer1991, pp. 585-616.

[2] K.J. Strnat, Modern permanent magnets for applications in electro-technology, Proceedings of the IEEE, 78 (1990) 923-946.
http://dx.doi.org/10.1109/5.56908

[3] http://www.magneticsmagazine.com/main/news/permanent-magnet-market-will-reach-28-70-billion-in-2019/.

[4] http://www.marketsandmarkets.com/Market-Reports/permanent-magnet-market-806.html.

[5] O. Gutfleisch, M.A. Willard, E. Brück, C.H. Chen, S. Sankar, J.P. Liu, Magnetic materials and devices for the 21st century: stronger, lighter, and more energy efficient, Advanced materials, 23 (2011) 821-842.
http://dx.doi.org/10.1002/adma.201002180

[6] M. Olszewski, Final Report on Assessment of Motor Technologies for Traction Drives of Hybrid and Electric Vehicles, Subcontract No. 4000080341, Oak Ridge National Laboratory, (2011).

[7] B.D. Cullity, C.D. Graham, Introduction to magnetic materials, John Wiley & Sons2011.

[8] K. Honda, Magnet-steel, in: U.S.P. OFFICE (Ed.), Google Patents, 1920, pp. 1-2.

[9] J. Becker, Rare-Earth-Compound Permanent Magnets, Journal of Applied Physics, 41 (1970) 1055-1064.
http://dx.doi.org/10.1063/1.1658811

[10] I. Al-Omari, R. Skomski, R. Thomas, D. Leslie-Pelecky, D.J. Sellmyer, High-temperature magnetic properties of mechanically alloyed SmCo 5 and YCo 5 magnets, Magnetics, IEEE Transactions on, 37 (2001) 2534-2536.

[11] A.L. Robinson, Powerful new magnet material found, in: K.A. Gschneidner Jr, J. Capellen (Eds.) 1787 - 19 87 Two Hundred Years of Rare Earths, North-Holland Publishing Co., Amsterdam1987, pp. 19-22.

[12] K. Kamino, Y. Kimura, T. Suzuki, Y. Itayama, Variation of the Magnetic Properties of Sm(Co,Cu)$_5$ Alloys with Temperature, Transactions of the Japan Institute of Metals, 14 (1973) 135-139.
http://dx.doi.org/10.2320/matertrans1960.14.135

[13]   S. Liu, A. Ray, H. Mildrum, Sintered $Sm_2(Co, Fe, Cu, Zr)_{17}$ magnets with light rare earth substitutions, Magnetics, IEEE Transactions on, 26 (1990) 1382-1384.

[14]   S. Liu, A. Ray, $Sm_2(Co, Fe, Cu, Zr)_{17}$ magnets with higher Fe content, Magnetics, IEEE Transactions on, 25 (1989) 3785-3787.

[15]   Y. Zhang, M. Corte-Real, G.C. Hadjipanayis, J. Liu, M.S. Walmer, K.M. Krishnan, Magnetic hardening studies in sintered $Sm(Co, Cu_x, Fe, Zr)_z$ 2: 17 high temperature magnets, Journal of Applied Physics, 87 (2000) 6722-6724.
http://dx.doi.org/10.1063/1.372820

[16]   J. Liu, Y. Zhang, D. Dimitrov, G. Hadjipanayis, Microstructure and high temperature magnetic properties of $Sm(Co, Cu, Fe, Zr)_z$ (z= 6.7–9.1) permanent magnets, Journal of applied physics, 85 (1999) 2800-2804.
http://dx.doi.org/10.1063/1.369597

[17]   J. Zhou, R. Skomski, C. Chen, G. Hadjipanayis, D.J. Sellmyer, Sm–Co–Cu–Ti high-temperature permanent magnets, Applied Physics Letters, 77 (2000) 1514-1516.
http://dx.doi.org/10.1063/1.1290719

[18]   J. Zhou, R. Skomski, D.J. Sellmyer, W. Tang, G.C. Hadjipanayis, Effect of Iron Substitution on the High-Temperature Properties of $Sm(Co, Cu, Ti)_z$ Permanent Magnets, MRS Proceedings, Cambridge Univ Press, 2001, pp. U2. 3.

[19]   J.J. Croat, Observation of large room-temperature coercivity in melt-spun $Nd_{0.4}Fe_{0.6}$, Applied Physics Letters, 39 (1981) 357-358.
http://dx.doi.org/10.1063/1.92728

[20]   J.J. Croat, J.F. Herbst, Melt-spun $R_{0.4}Fe_{0.6}$ alloys: Dependence of coercivity on quench rate, Journal of Applied Physics, 53 (1982) 2404-2406.
http://dx.doi.org/10.1063/1.330826

[21]   J.J. Croat, Magnetic hardening of Pr-Fe and Nd-Fe alloys by melt-spinning, Journal of Applied Physics, 53 (1982) 3161-3169.
http://dx.doi.org/10.1063/1.331014

[22]   N.C. Koon, B. Das, Magnetic properties of amorphous and crystallized $(Fe_{0.82}B_{0.18})_{0.9}Tb_{0.05}La_{0.05}$, Applied Physics Letters, 39 (1981) 840-842.
http://dx.doi.org/10.1063/1.92578

[23]   N.C. Koon, B. Das, J. Geohegan, Composition dependence of the coercive force and microstructure of crystallized amorphous $(Fe_xB_{1-x})_{0.9}Tb_{0.05}La_{0.05}$ alloys, Magnetics, IEEE Transactions on, 18 (1982) 1448-1450.

[24] G.C. Hadjipanayis, R. Hazelton, K. Lawless, New iron-rare-earth based permanent magnet materials, Applied Physics Letters, 43 (1983) 797-799. *http://dx.doi.org/10.1063/1.94459*

[25] M. Sagawa, S. Fujimura, N. Togawa, H. Yamamoto, Y. Matsuura, New material for permanent magnets on a base of Nd and Fe, Journal of Applied Physics, 55 (1984) 2083-2087. *http://dx.doi.org/10.1063/1.333572*

[26] J.J. Croat, J.F. Herbst, R.W. Lee, F.E. Pinkerton, High-energy product Nd-Fe-B permanent magnets, Applied Physics Letters, 44 (1984) 148-149. *http://dx.doi.org/10.1063/1.94584*

[27] J.J. Croat, J.F. Herbst, R.W. Lee, F.E. Pinkerton, Pr-Fe and Nd-Fe-based materials: A new class of high-performance permanent magnets, Journal of Applied Physics, 55 (1984) 2078-2082. *http://dx.doi.org/10.1063/1.333571*

[28] D. Brown, B.-M. Ma, Z. Chen, Developments in the processing and properties of NdFeB-type permanent magnets, Journal of Magnetism and Magnetic Materials, 248 (2002) 432-440. *http://dx.doi.org/10.1016/S0304-8853(02)00334-7*

[29] M. Kramer, R. McCallum, I. Anderson, S. Constantinides, Prospects for non-rare earth permanent magnets for traction motors and generators, JOM, 64 (2012) 752-763. *http://dx.doi.org/10.1007/s11837-012-0351-z*

[30] http://www.grandviewresearch.com/industry-analysis/permanent-magnet-industry.

CHAPTER 3

Properties and Synthesis of Hexaferrites

S.H. Mahmood

Physics Department, The University of Jordan, Amman, Jordan

s.mahmood@ju.edu.jo

Abstract

Hexaferrites belong to an extremely important class of magnetic oxides, which exhibit properties suitable for a wide range of applications. In particular, M-type hexaferrites is cost effective for permanent magnet applications. Consequently, the synthesis and improvement of the properties of these ferrites have attracted considerable interest, and different synthesis techniques of M-type powders were developed. Chemical synthesis routes were adopted to improve phase homogeneity and magnetic properties of the final product. In this chapter, we review the main structural and magnetic characteristics of this important class of materials, and briefly discuss some of the important methods utilized for their production.

Keywords

M-Type Hexaferrite; Structural Properties; Magnetic Properties; Synthesis

Contents

1. INTRODUCTION

Hexaferrites (also known as hexagonal ferrites) is the term used for a family of magnetic oxides containing iron as a major component, and having hexagonal structure. These materials were discovered and described by the Philips group in the early 1950s [1-5]. Due to their easy production, corrosion resistance, high electrical resistance in comparison with metallic magnetic materials, and low cost of production, these materials received an exponentially increasing interest by scientists as well as engineers and technologists. Fig. 1 demonstrates the exponential growth of the annual average of the number of scientific articles published so far during the decades following the discovery of hexagonal ferrites.

Hexaferrites are found in a wide range of industrial and technological applications; in auto industry, telecommunication, consumer electronics, data processing and information technology, instrumentation, etc. [6-9]. The number of motors commonly used in a typical household has grown orders of magnitudes in the last few decades. With hundreds of millions of motors and actuators produced annually, the cost-effectiveness of magnet production becomes a critical factor. The importance of these magnetic materials for commercialization may be indicated by the exponential increase of the annual average of patents registered during the decades following their discovery (Fig. 2).

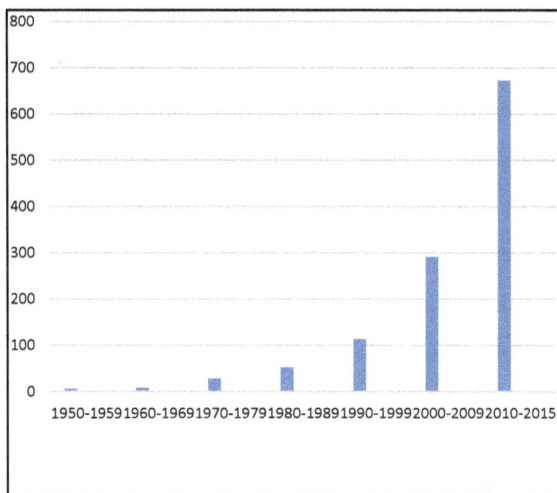

Fig. 1: Annual average of the number of hexaferrite articles published during the indicated periods (from Google Scholar).

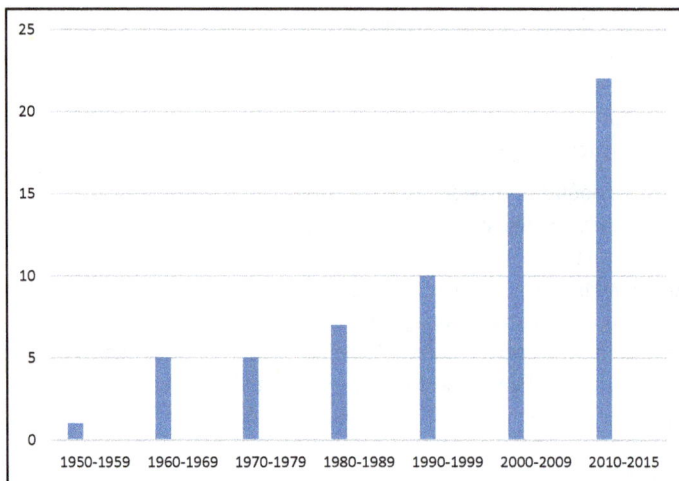

Fig. 2: Annual average of the number of hexaferrite patents during the indicated periods (from Google Scholar).

The different types of hexaferrites have different magnetic properties depending on their crystallographic and magnetic structure. These ferrites are characterized by a high magnetocrystalline anisotropy. The coercivity of single magnetic domain crystals of different types of hexaferrites range from few tens or hundreds of Oerteds to few thousands of Oersteds, covering the range for a wide band of applications. Of these types, M-type barium hexaferrite is characterized by a relatively high uniaxial magnetocrystalline anisotropy, resulting in a relatively high coercivity suitable for PM applications. The growing demand for PM applications, and the market demand for cheaper magnets resulted in the dominance of hexaferrites in the PM market, where BaM alone accounts for 50% of the PM market volume [8]. Hexaferrites are generally insulating oxides with low eddy current losses, and are characterized by high permeability and moderate permittivity in the frequency range from dc to 300 GHz, which makes them unparalleled for a wide range of electronic applications [9]. Since the main concern of this book is permanent magnetic materials, the forthcoming sections will focus on the M-type hexaferrite, which is the material mainly used for the production of ferrite PMs today.

2.   CRYSTAL STRUCTURE

The basic structure of all hexaferrites is built from oxygen hexagonal close-packed layers, with oxygen ions replaced substitutionally by large divalent ions (Ba^{2+}, Sr^{2+}, Pb^{2+}) in some of the oxygen layers. This substitution is possible due to the fact that the ionic radii of the divalent cations (1.43 Å for Ba^{2+}, 1.27 Å for Sr^{2+}, and 1.32 Å for Pb^{2+}) are comparable to that of O^{2-} (1.32 Å). Different stacking sequences of hexagonal close-packed layers give different crystal symmetries. The ABABA… stacking sequence results in a hexagonal close-packed structure, whereas the ABCABC… stacking sequence gives a cubic (spinel) close-packed structure. In both cases, the distance between the centers of two layers on top of each other is $\sqrt{(8/3)}r$, where r is the radius of the ion in the layer (O^{2-}). In the close-packed structures discussed above, this distance is about 2.30 Å. Due to the larger ionic radius of Ba^{2+}, however, the distance between a layer containing a Ba ion and the next oxygen layer (without Ba) is slightly higher (2.35 Å), whereas the distance between two successive layers each containing a Ba ion is 2.40 Å [3]. This can be used as a rule of thumb to determine the length of the unit cell along the hexagonal c-axis.

The way by which the large cation substitutes oxygen anions in the hexaferrite structure leads to the formation of three basic structural blocks: The S block composed of two oxygen layers, the R block composed of three oxygen layers with one oxygen anion replaced by the large divalent cation at the central layer, and the T block composed of four oxygen layers with one oxygen anion replaced by one large cation at each of the two central layers. The structure of the hexaferrites contains small metal ions located at the interstitial sites between the layers.

The sequence of stacking of these blocks results in the formation of the different types of hexaferrites, namely, the M, Y, W, Z, X, and U types. These types of hexaferrites can also be visualized as being constructed from combinations of the M-type, the Y-type, and the S block. The crystal structures of all these compounds are closely related, complex hexagonal structures. The unit cell dimension for the different compounds vary mainly along the hexagonal c-axis, whereas their lattice parameter a is similar (about 5.88 Å [3]. Further, the types of the small metal ions at the interstitial sites of the hexaferrite structure has a small effect on the lattice parameters of a given hexaferrite type. The chemical formulas and stacking sequences of these types of hexaferrites are listed in Table 1.

Table 1: Chemical composition and stacking of the structural blocks in the different types of hexaferrites[10].

Type	Combination	Molecular formula	Structural stacking
M	M	$BaFe_{12}O_{19}$	$RSR*S*$
Y	Y	$Ba_2Me_2Fe_{12}O_{22}$	$TSTSTS$
W	M+S	$BaCo_2Fe_{16}O_{27}$	$RSSR*S*S*$
Z	M+Y	$Ba_3Co_2Fe_{24}O_{41}$	$RSTSR*S*T*S*$
X	2M+S	$Ba_2Co_2Fe_{28}O_{46}$	$RSR*S*S*$
U	2M+Y	$Ba_4Co_2Fe_{36}O_{60}$	$RSR*S*T*S*$

The first discovered hexaferrite material is the barium M-type (BaM) with unit cell built from the sequential stacking of the R and S blocks $RSR*S^*$, where the star indicates rotation of the block by 180° about the hexagonal c-axis. The chemical composition of the S block is $[Fe_6^{3+}O_8^{2-}]^{2+}$, while the composition of the R block is $[BaFe_6^{3+}O_{11}^{2-}]^{2-}$. The unit cell of BaM is therefore chemically neutral, and contains two molecular formula units with composition $BaFe_{12}O_{19}$, having a molecular mass of 1111.48 g/mol. In the M-type hexaferrite structure, there are five distinct crystallographic interstitial sites occupied by the small metal ions, one tetrahedral ($4f_1$) site, three octahedral ($2a, 4f_2, 12k$) sites, and one five-coordinated bi-pyramidal (trigonal) $2b$ site. The $4f_1$ and $2a$ sites are located within the S block, the $4f_2$ and $2b$ sites within the R-block, and the $12k$ site at the R-S interfaces. The R and S structural blocks of BaM are shown in Fig. 3 and 4.

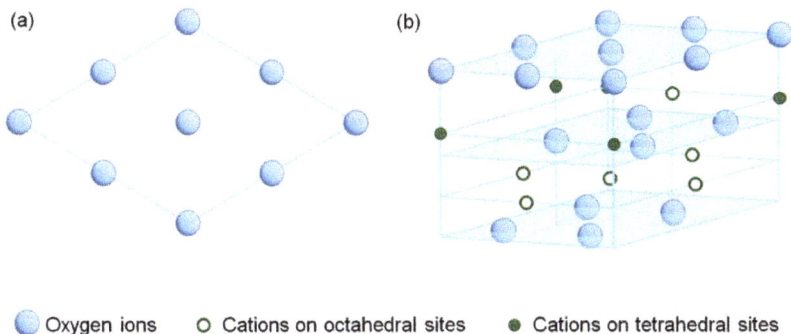

Fig. 3: The S block of the hexaferrite structure. (a) shows the oxygen layer in the R block on top of the S block [11].

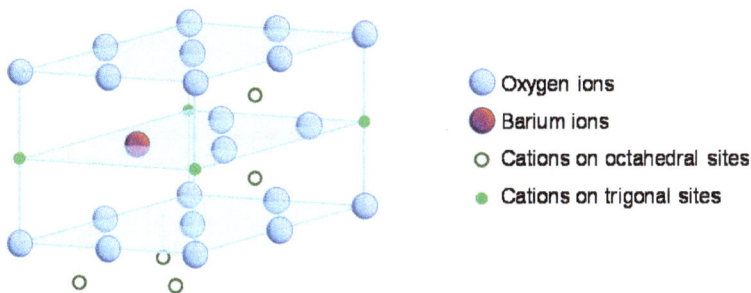

Fig. 4: The R block of the hexaferrite structure [11].

The unit cell of BaM structure consists of 10 hexagonal layers with the stacking sequence as in Fig. 5. In the figure, the Ba ions in R-block central layers are indicated by solid spheres. Notice that the stacking sequence of the layers reveals two distinct crystal symmetries, namely, cubic ABCABC stacking type (in the S block), and hexagonal ABAB stacking type (in the R block). The typical lattice parameters of the BaM hexaferrite are: $a = b = 5.88$ Å and $c = 23.2$ Å, and its x-ray density is about 5.28 g/cm^3 [3].

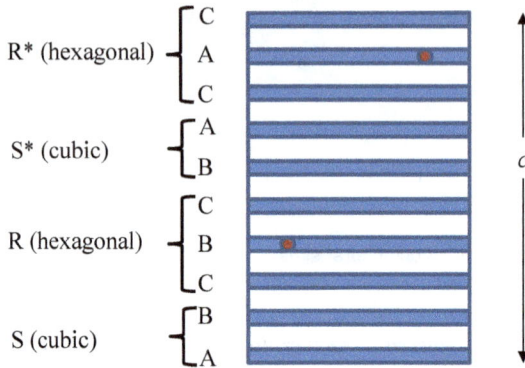

Fig. 5: The stacking sequence of the hexagonal close-packed layers in the unit cell of BaM hexaferrite structure, indicating the structural R and S blocks, and the Ba ions (solid spheres) at the central layer of the R block.

X-ray diffraction (XRD) pattern of a typical isotropic BaM hexaferrite cylindrical disk magnet produced in our lab is shown in Fig. 6, together with the standard pattern (JCPDS: 00-043-0002) for BaM hexaferrite. Miller indices of the major reflections are indicated in the figure. The sample was prepared by ball milling the appropriate amounts of high purity α-Fe$_2$O$_3$ and BaCO$_3$ precursor powders, pressing the milled powder mixture under high pressure, and sintering the resulting disk at 1150° C for 2 hours. Structural refinement of the pattern using Rietveld analysis revealed the presence of a single hexagonal BaM phase with space group $P6_3/mmc$, and no secondary phases. The two mirror planes in this structure are the planes containing the layers with Ba^{2+} ions in the R-block. Structural refinement parameters obtained by Rietveld analysis ($a = 5.89$ Å; $c = 23.21$ Å) were consistent with reported lattice parameters of BaM in the literature [6].

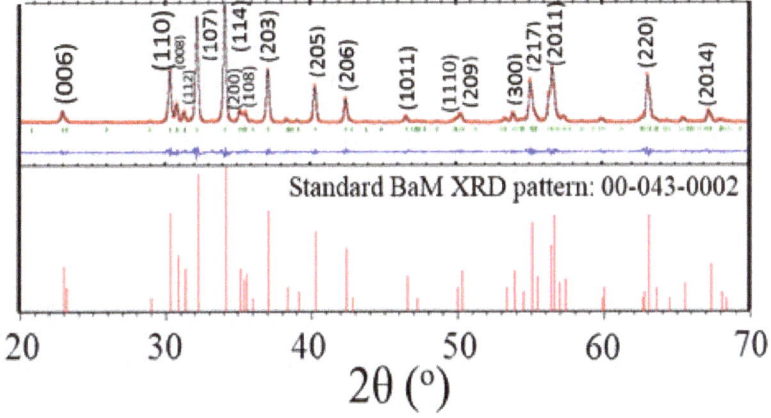

Fig. 6: XRD pattern of a single-phase BaM sample prepared in our lab.

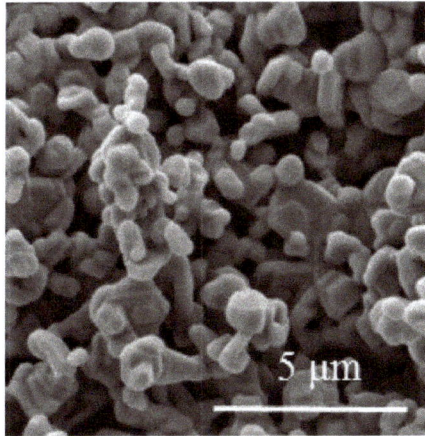

Fig. 7: SEM image of BaM hexaferrite prepared in our lab by ball milling and solid state reaction.

Scanning electron microscopy (SEM) imaging of the sample revealed a porous, granular structure composed of irregular particle shapes of typical size below 0.5 μm (Fig. 7).

Platelet-like particles with approximately hexagonal shape can also be identified in the image. The particle shape and size is influenced by the sintering temperature, where higher temperatures generally leads to particle growth, and results in the deterioration of the magnetic properties of the magnetic material. Therefore, the material should be synthesized under controlled experimental conditions to obtain particle morphology and size suitable for practical applications.

The main structural features of the commercially important SrM hexaferrite are similar to those of BaM. However, due to the fact that the ionic radius of Sr is smaller than that of Ba, the lattice parameters of SrM are somewhat smaller, where $a = 5.86$ Å; $c = 23.03$ Å [8].

Barium Y-type hexaferrite (Me$_2$Y) has a chemical formula Ba$_2$Me$_2$Fe$_{12}$O$_{22}$, where Me is a divalent metal ion such as Co^{2+}, Zn^{2+}, Ni^{2+}, Mg^{2+}, Mn^{2+}. This material is characterized by a c-plane easy axis at room temperature, and is magnetically soft in comparison with M-type hexaferrite. Therefore, Y-type hexaferrites are not of interest to the PM industry. However, this compound has a multitude of other applications [6], which are beyond the scope of this book, but justifies the brief discussion of some of its characteristics.

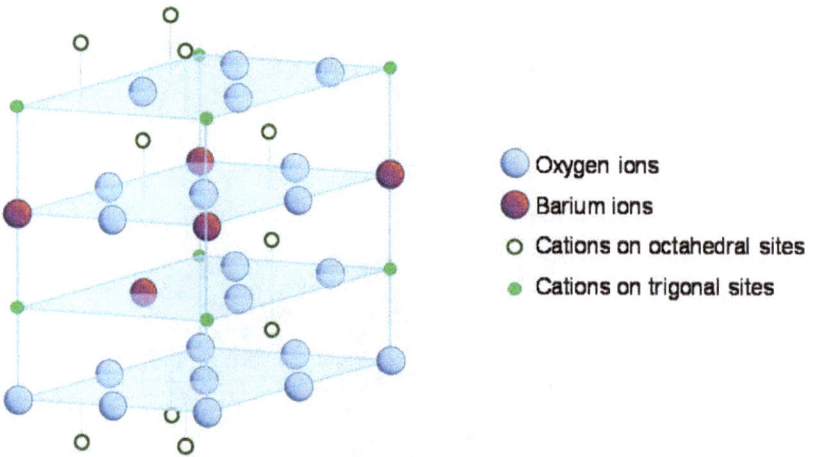

Fig. 8: The T block in a hexaferrite structure

The unit cell is built from chemically neutral S and T blocks with the structural stacking STS′T′S″T″, where the primes and double primes indicate rotations of the blocks by 120°

about the hexagonal c-axis. The T structural block in the Y-type hexaferrite is shown in Fig. 8 [12].

The chemical composition of the S block is $[Me_2^{2+}Fe_4^{3+}O_8^{2-}]^0$, while that of the T block is $[Ba_2^{2+}Fe_8^{3+}O_{14}^{2-}]^0$. Accordingly, the unit cell contains 3 molecular formula units with composition $Ba_2Me_2Fe_{12}O_{22}$. The unit cell is composed of 18 hexagonal layers, with the small metal ions (Me^{2+} and Fe^{3+}) occupying 6 different tetrahedral ($6c_{IV}, 6c_{IV}^*$) and octahedral ($3a_{VI}, 6c_{VI}, 3b_{VI}, 18h_{VI}$) interstitial sites. The $3a_{VI}$ and $6c_{IV}$ sites are located within the S block, the $3b_{VI}, 6c_{IV}^*$ and $6c_{VI}$ within the T block, and the $18h_{VI}$ site at the S-T interfaces. The exact lattice parameters of Y-type hexaferrite structure vary slightly from their typical values of $a = b = 5.88$ Å and $c = 43.56$ Å, depending on the type of Me ion [3]. For Co_2Y, the molecular mass is 1410 g, and the x-ray density is 5.40 g/cm^3 [8].

X-ray diffraction pattern of a typical Co_2Y sample prepared in our lab is shown in Fig. 9. The sample was prepared by ball milling appropriate amounts of high-purity precursor powders, compacting at high pressure, and sintering at 1200° C for 2 hours. Rietveld refinement of the pattern revealed a single hexagonal phase consistent with the standard pattern (JCPDS: 00-044-0206) with space group R_3m. The lattice parameters of this compound are: $a = b = 5.87$ Å and $c = 43.54$ Å.

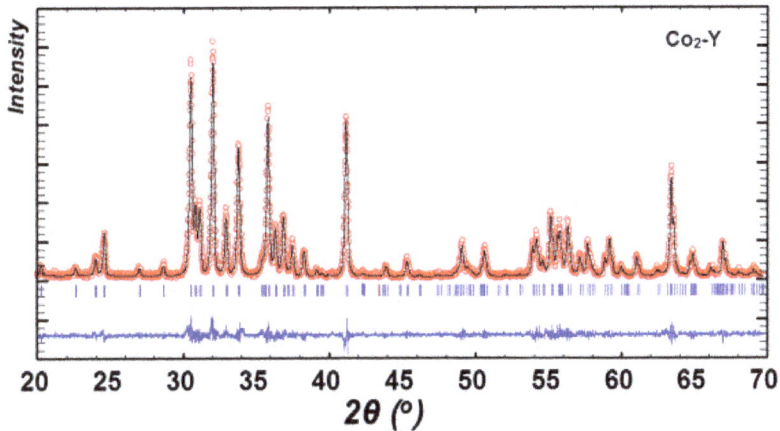

Fig. 9: X-ray diffraction pattern of Co₂Y hexaferrite compound.

SEM image of the sample is shown in Fig. 10. The sample is composed of porous, granular assembly of interconnected platelet-like particles. The hexagonal symmetry of the particles is evident. The typical particle size ranges from 1μm to few μm in the in-plane dimension.

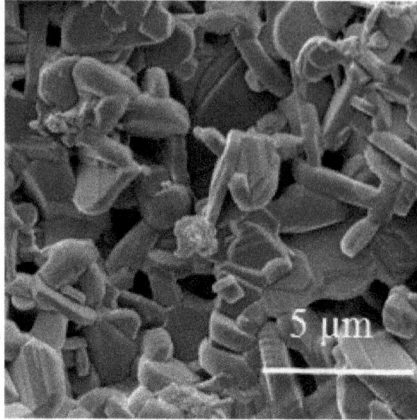

Fig. 10: SEM image of Co$_2$Y hexaferrite prepared in our lab by ball milling and solid state reaction.

3. MAGNETIC PROPERTIES

Magnetic ions in M-type hexaferrites occupy five different crystallographic sites $(2a, 4f_1, 4f_2, 2b, 12k)$, three of which are octahedral, one is tetrahedral, and one is five-coordinated bi-pyramidal site as mentioned earlier. Superexchange interactions between these ions via the intervening O^{2-} ions give rise to the magnetic structure in M-type hexaferrites, and the strengths of these interactions determine their unique magnetic properties.

Magnetic ions in the hexaferrite lattice are too far to interact via the normal exchange forces between localized magnetic moments. Accordingly, indirect superexchange interactions between magnetic ions are dominant. If two magnetic ions (M_1 and M_2) interact via an intervening O^{2-} ion, the strength of the interaction is determined by the M_1-O-M_2 bond angle, being high at bond angles near 180°, and rather weak near 90° as demonstrated by Fig. 11. According to the figure, spin-up moment of M_1 localized near an O^{2-} repels the spin-up orbit of the O^{2-} ion, resulting in a spatial separation of its spin-up and spin-down orbits. If the bond angle is close to 180°, superexchange interactions force the magnetic moment of M_2 ion to be antiparallel to the closest spin orbit of the O^{2-} ion, and consequently, antiparallel to the orientation of the magnetic moment of M_1, resulting in antiferromagnetic coupling between the two ions. On the other hand, the extreme case of a 90° bond angle does not force antiparallel alignment of the magnetic moments of M_1 and M_2.

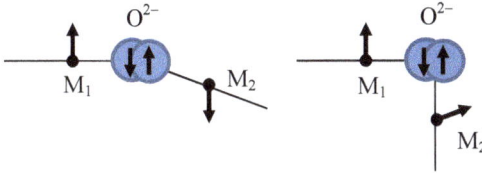

Fig. 11: Schematic diagram of the superexchange interaction forcing antiparallel alignment of magnetic moments in the case of M-O-M bond angle close to 180°.

Considering magnetic interactions between two neighboring magnetic sublattices A and B, the M-O-M bond angle between two magnetic ions in the same sublattice (A–A or B–B type interactions) is usually close to 90°. For example, the bond angle of two $4f_2$ ions in BaM hexaferrite($<4f_2 - O - 4f_2 >$) is 83.8°, and the bond angles of $<12k - O - 12k>$ bonds fall in the range 88.2°–97.8° [13]. This would lead to a significantly weaker intra-sublattice interactions in comparison with inter-sublattice A–B type interactions, where the bond angle $<12k - O - 4f_2>$ is 128.1°. Accordingly, superexchange interactions dominated by inter-sublattice interactions. Fig. 12 illustrates that a magnetic ion at $2b$ site is coupled through superexchange interactions to the two $4f_2$ ions in the R block via the same (O3) ion, making angles of 132.9° and 143.3°. This is by far the strongest superexchange interaction in the hexaferrite lattice, and consequently, the $2b - 4f_2$ is the orienting interaction forcing antiparallel coupling between $2b$ (spin-up) and $4f_2$ (spin-down) sites. On the other hand, the $12k - 4f_2$ interaction is stronger than the $2b - 12k$ interaction as can be deduced from bond lengths and bond angles presented in Table 2 [13], forcing the $12k$ to be a spin-up sublattice. Further, the $12k - 4f_1$ and $2a - 4f_1$ interactions are rather stronger than the $2a - 12k$ interactions, forcing the $4f_1$ to form a spin-down sublattice, and the $2a$ to form a spin-up sublattice.

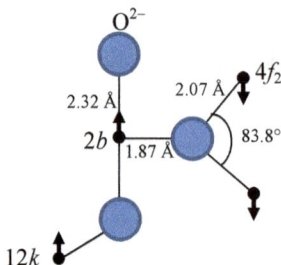

Fig. 12: Schematic diagram of the strong orienting $2b - 4f_2$ superexchange interactions.

Table 2: Interatomic distances (Å) and bond angles (°) in BaM hexaferrite. Values corresponding to bonding with O1 (4e), O2 (4f), O3 (6h), O4 (12k) and O5 (12k) are presented [13].

Distances (Å)		Bond Angles (°)	
$2a - O_4$	2.000	$2a - O_4 - 4f_1$	126.3
$2b - O_1$	2.128	$2a - O_4 - 12k$	95.4
$2b - O_1$	2.468	$2b - O_1 - 12k$	119.9
$2b - O_3$	1.867	$2b - O_3 - 4f_2$	132.9
$4f_1 - O_2$	1.894	$2b - O_3 - 4f_2$	143.3
$4f_1 - O_4$	1.894	$12k - O_2 - 4f_1$	126.5
$4f_2 - O_3$	2.073	$12k - O_4 - 4f_1$	121.3
$4f_2 - O_5$	1.969	$4f_2 - O_3 - 4f_2$	83.8
$12k - O_1$	1.985	$4f_2 - O_5 - 12k$	128.1
$12k - O_2$	2.092	$12k - O_1 - 12k$	97.3
$12k - O_4$	2.114	$12k - O_2 - 12k$	88.2
$12k - O_5$	1.932	$12k - O_4 - 12k$	89.7
		$12k - O_5 - 12k$	97.8

The relative strengths of the superexchange interactions discussed above are responsible for the magnetic structure in M-type hexaferrites. At zero temperature, thermal effects are negligible, and superexchange interactions in M-type hexaferrites result in a collinear magnetic structure consistent with the Gorter model [14]. Spin orientations, coordinations, number of magnetic ions per molecule, and the positions of the five sublattices in the M-type hexaferrite lattice are as shown in Table 3. According to this spin structure, and taking into consideration that the magnetic moment of an Fe^{3+} ion is 5 μ_B, the magnetic moment per molecule is: $[1\ (2a) + 1\ (2b) - 2\ (4f_1) - 2\ (4f_2) + 6\ (12k)] \times 5$ $\mu_B = 20\ \mu_B$. This molecular moment corresponds to a saturation magnetization of about 100 emu/g at 0 K, which is consistent with experimental results [15].

Table 2: Spin orientation and coordination of the five sublattices in M-type hexaferrite.

Sublattice	Coordination	Block	Ions/molecule	Spin orientation
2a	Octahedral	S	1	up
$4f_1$	Tetrahedral	S	2	down
$4f_2$	Tetrahedral	R	2	down
2b	Bi-pyramidal	R	1	up
12k	Octahedral	R–S	6	up

The saturation magnetization decreases with increasing the temperature due to thermal effects. At room temperature, the saturation magnetization is only 72 emu/g. Unlike metallic magnets such as Fe, Co, and Ni, the temperature dependence of the saturation magnetization is almost linear. This is due to the different temperature dependence of the magnetizations of the five sublattices as depicted by the schematic diagram in Fig. 13. The fast drop of the $12k$ magnetization with temperature is responsible for this behavior. This fast drop is due to the competing $2b - 12k$ interaction, which is relatively strong compared to interactions of ions in other sublattices with neighboring ions having parallel spins.

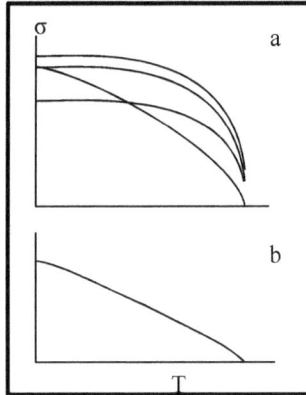

Fig. 13: Magnetization, σ, vs temperature for (a) the different sublattices and (b) the net magnetization of M-type hexaferrite material.

The Curie temperature for BaM is 450° C, and that of SrM is slightly higher (460° C). M-type hexaferrites are characterized by a high uniaxial magnetocrystalline anisotropy field which is related to the first anisotropy constant by [16]:

$$H_a = \frac{2K_1}{M_s} \tag{1}$$

Substituting the values $K_1 = 3.3 \times 10^6$ erg/cm^3 and $M_s = 380$ emu/cm^3 for BaM at room temperature, the magnetocrystalline anisotropy field for this hexaferrite is 17.4 kOe. According to Stoner – Wohlfarth model for a random assembly of single domain BaM hexaferrite particles, the theoretical coercivity is given by:

$$H_c = 0.48 \times \frac{2K_1}{M_s} \tag{2}$$

Accordingly, the theoretical coercivity is 8350 Oe, which is higher than any experimentally observed value (normally 2000 – 4000 Oe) [17-20]. For thin platelet-like particles, however, shape anisotropy may play an important role in determining the sample coercivity. If the easy axis for shape anisotropy is in the plane of the plates, and the magnetocrystalline anisotropy easy axis is perpendicular to the plane, the two anisotropies oppose each other, and the resulting coercivity is determined by the relation [21]:

$$H_c = 0.48 \left(\frac{2K_1}{M_s} - 4\pi M_s \right) \tag{3}$$

In a recent study [20], the magnetocrystalline anisotropy field of BaM prepared by the standard ceramic method was reported to be about 12 kOe, and the saturation magnetization was about 66 emu/g. Eq. (3) predicts that the coercivity of the sample should be about 3660 Oe, which is in reasonable agreement with the experimental coercivity of about 4000 Oe. The lower value of the theoretical coercivity could be associated with the fact that the demagnetizing coefficient is lower than the 4π value in Eq. (3). This is an indication that the demagnetizing coefficient along the perpendicular axis is not negligible, resulting in a lower effective shape anisotropy than indicated by Eq. (3).

In addition to the particle shape, the particle size of the magnetic powder plays an essential role in determining the coercivity of the sample. For a typical superparamagnetic sample with particle size below the superparamagnetic critical size (D_s), the coercivity is identically zero. As the particle size increases above D_s, the coercivity increases up to the critical single domain particle size (D_c), and then starts decreasing as a consequence of the development of multi-domains within the particles as demonstrated by the schematic diagram in Fig. 14.

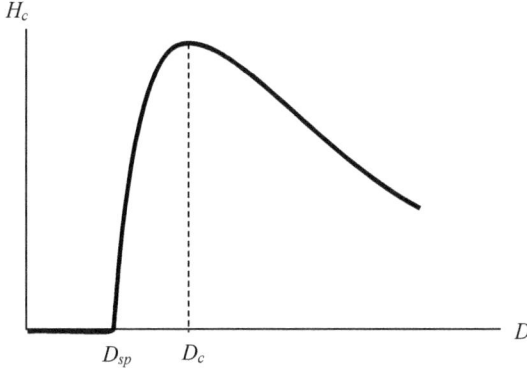

Fig. 14: Coercivity of a powder magnetic material as a function of particle size.

The behavior of the coercivity with increasing particle size was confirmed by a systematic investigation of the coercivity of a set of BaM samples prepared by the conventional ceramic method, and sintered at different temperatures between 1000° C and 1400° C [22]. The coercivity was found to drop from about 4 kOe for the sample sintered at 1000° C (particle size of about 0.55 μm), to few Oe for samples sintered at > 1300° C, with an average particle size order of magnitude larger. The results of this study confirm that the critical single domain size in BaM is about 0.5μm, which is consistent with the theoretical critical size of 460 nm.

M-type hexaferrites are characterized by a linear $B–H$ curve far into the second quadrant of the hysteresis loop. For an ideal, densified anisotropic magnet, the remanent magnetization is equal to the saturation magnetization (B_r = 4777 G for BaM at room temperature). Also, the coercivity of such a magnet is $H_{cB} = B_r$, and the theoretical maximum energy product is 5.7 MGOe (45 kJ/m^3). In practice, this value cannot be achieved since it is technically impossible to produce a magnet with the theoretical density, and perfectly aligned with a squareness ratio of 1. The magnetic properties of the ferrite magnet can be improved by adopting appropriate scenarios to improve the saturation magnetization, coercivity, and squareness ratio. Using Sr instead of Ba was found to result in improving the magnetic properties of the magnet. Anisotropic ferrite magnets with energy products as high as 35 kJ/m^3 were commercially produced.

Fig. 15 shows the *B–H* hysteresis loop, together with the *J–H* loop ($J = 4\pi M$) for a typical isotropic BaM magnet produced in lab using standard solid state reaction. The figure shows a perfectly linear *B–H* relation in the second quadrant of the hysteresis loop. The remanent induction for this magnet $B_r = 2240$ G, and the coercivity $H_{cB} = 1.825$ kOe (145 kA/m) and $H_{cJ} = 4.02$ kOe (320 kA/m). Fig. 16 indicates that the maximum energy product of this magnet is 1.03 MGOe (8.2 kJ/m^3). On the other hand, the magnetic properties of an isotropic SrM magnet (hysteresis loops shown in Fig. 17) prepared under identical conditions are: $B_r = 2320$ G, $H_{cB} = 1.9$ kOe (151 kA/m) and $H_{cJ} = 4.5$ kOe (358 kA/m). According to Fig. 18, the maximum energy product of this magnet is $(BH)_{max} = 1.12$ MGOe (9 kJ/m^3).

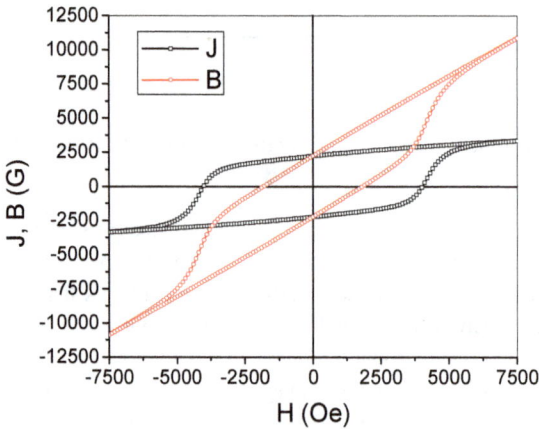

Fig. 15: Magnetization, (J) and induction (B) loops of isotropic BaM magnet.

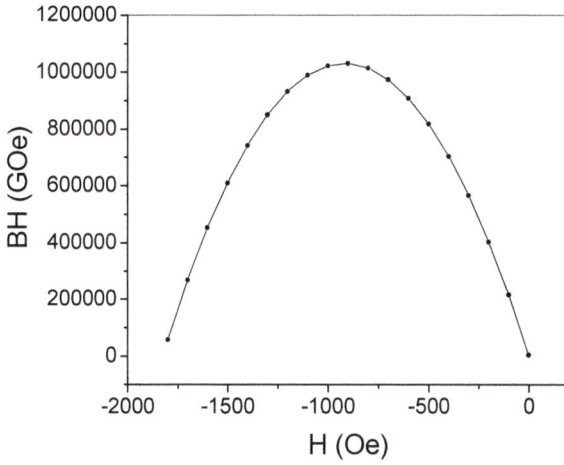

Fig. 16: Energy product vs. H for the isotropic BaM magnet.

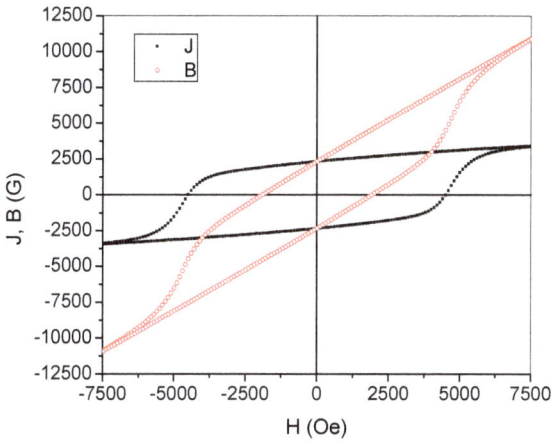

Fig. 17: Magnetization, (J) and induction (B) loops of isotropic SrM magnet.

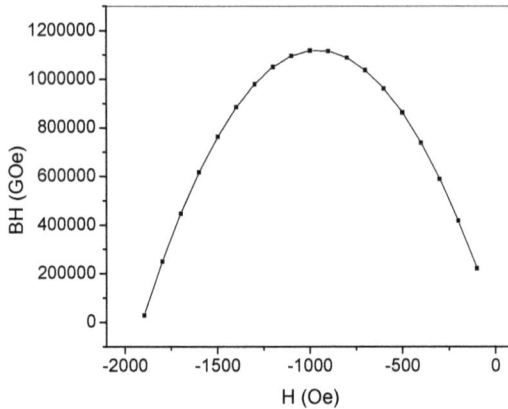

Fig. 18: Energy product vs. H for the isotropic SrM magnet.

4. PROCESSING OF FERRITE MAGNETS

Since M-type hexaferrite is commercially the most important material, we focus our discussion on the production of BaM and SrM ferrite magnets. Ferrite magnets are produced in two major categories: sintered magnets, and bonded magnets. Bonded magnets have the advantage of flexibility and easy cutting and machining into desired shapes compared to sintered magnets. However, bonded magnets have lower residual induction and energy products, which limits their use to applications not requiring high magnetic properties, such as magnetic seal strips and toys. Sintered ferrite magnets, with higher magnetic properties, are mainly used in simple motors for household appliances and the auto-industry.

Ferrite magnets are commercially processed by powder-metallurgy. The initial powders, composed of analytic grade α-Fe_2O_3 and barium or strontium carbonates or oxides, are mixed and milled down to micron-size particles for powder homogenization. The powders are then roasted at temperatures in excess of $1100°$ C, where solid state reaction leads to the synthesis of the desired M-type hexaferrite material. For the production of isotropic bonded magnets, the hexaferrite powder is mixed with a binder, such as epoxy or rubber, which is subsequently cured by an appropriate heat treatment, and then formed into a desired shape. Anisotropic bonded magnets with better magnetic properties can be produced by magnetic alignment of the particles during the magnetic production stages.

For the production of sintered isotropic ferrite magnets, the hexaferrite powder is molded into a desired shape by high pressure compaction, and then sintered at temperatures ~ 1200° C. Alternatively, the starting powders are compacted into desired shapes, and sintered at ~ 1200°. The final processing of the magnet involves surface grinding and smoothing, and magnetization in an applied magnetic field. Fig. 18 is a flowchart illustrating the production stages of sintered ferrite magnets. The remanence magnetization of isotropic magnets fabricated from powders composed of single domain, randomly oriented particles is typically half the saturation magnetization, resulting in an energy product typically around 1 MGOe (8 kJ/m^3).

The remanent flux density can be doubled, and the energy product enhanced four times by aligning the ferrite particles in a magnetic field prior to the final production of the magnet. Such magnets with oriented particles were produced as early as 1954 [23]. This can be achieved by pressing BaM particle suspension in a magnetic field, drying, and sintering the compacts at a temperature ~ 1200° C [24]. A systematic investigation of the magnetic properties of BaM ferrites prepared from nonstoichiometric ratios of Fe:Ba and sintered at temperatures from 1180° C − 1280° C resulted in anisotropic magnets with optimal magnetic properties of B_r = 0.388 T, H_{cB} = 2100 Oe, and $(BH)_{max}$ = 3.52 MGOe; this was achieved at an Fe:Ba molar ratio of about 11.0 (15% wt of BaO) [24]. The energy product of the best magnet obtained was still significantly lower than the theoretical value of 5.7 MGOe, which could be due to the occurrence of particles larger than the single domain critical size, resulting in a degradation of the magnetic properties due to domain walls within the particles. To produce an anisotropic magnet with the highest possible coercivity, the particle size should be close to, and below the critical single domain size of about 0.5 µm.

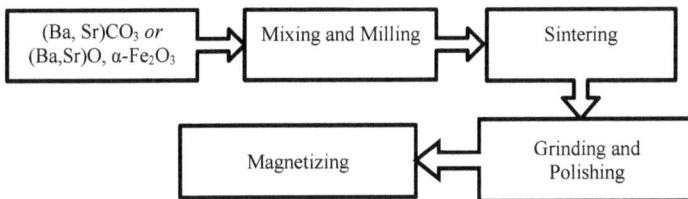

Fig. 18: Flowchart illustrating the processing of isotropic sintered ferrite magnet.

5.   SYNTHESIS OF FERRITE POWDERS

Although the conventional ceramic method discussed above is widely used for commercial production of hexaferrite powder, several other techniques involving chemical routes were adopted for the production of the magnet powders. The products of the various techniques revealed a wide range of magnetic properties depending on the purity and homogeneity of the ferrite phase, as well as on the shape and size of the particles in the powder, and the particle size distribution. The experimental conditions adopted for the production of ferrites by a given method were found to be crucial in determining the magnetic properties of the final product. These factors include heat treatment, chemicals used in the synthesis, stoichiometry of the precursor powder, and other relevant experimental conditions.

In the following presentation we briefly discuss main techniques used to prepare ferrite precursor powders. Further details of variations of the synthesis techniques are found in an excellent recently published review article [8].

5.1   CONVENTIONAL CERAMIC METHOD

The standard ceramic method of preparing M-type hexaferrites by mixing, finely grinding, and sintering the appropriate amounts of metal oxides and carbonates precursor powders is a simple method which is widely used in the production of ferrite magnetic materials [15, 25-29]. Usually, appropriate molar ratios of high purity hematite (α-Fe$_2$O$_3$) and divalent metal (Ba or Sr) carbonate are mixed and dry-milled or wet-milled in hardened stainless steel vials (or in zirconia, tungsten carbide, or other mechanically hard materials), using balls (typically 1 cm in diameter) made of the same material of the vials. In wet milling, the suspension of the particles in the liquid medium (water, alcohol, or acetone for example) makes the milling more efficient than dry milling. The ball-powder-liquid mass ratios are selected to give optimum results. Typical ball-powder mass ratios are in the range 8 – 14, while powder to liquid ratio is about 1:1. The high energy ball milling is performed at few hundred cps for periods dictated by the requirement of the final particle size; typical milling times are 10 – 20 h. The muddy product is dried, and the resulting powder is compacted into desired shape at high pressure for near full densification. The compacted magnets are then sintered at high temperature (900° C – 1300° C) for few hours. Compaction of wet-milled BaM powder at a pressure of ~ 250 MPa, and sintering at 1220° C for 3 h resulted in a density of the pellets of about 90% of the theoretical density [30], while compaction at ~ 500 MPa and sintering at 1100° C for 2 h was reported to result in ~ 80 – 93 % densification [31]. On the other hand, pellets compacted at 30 MPa and sintered at 1150° C were reported to have a density below 70% of the theoretical density [32].

The typical Fe:Ba ratio used for the production of BaM hexaferrite powder is normally 11 – 12. However, some studies were devoted to the investigation of the effect of Fe:Ba ratio on the hexaferrite phase evolution and properties of the powder products [11, 12, 33]. In order to determine the optimal Fe:Ba ratio in ball-milled powders sintered at 1100° C, a detailed analysis of the structural phases in a series of BaM hexaferrites with Fe:Ba ratios ranging from 11.5 – 16.16 was carried out. The results indicated the coexistence of BaM and α-Fe_2O_3 phases in samples with theoretical or higher Fe:Ba ratios. The analysis revealed that the optimal ratio for the production of a single BaM phase was 11.7 [11].

The structural phases in Ba-rich samples prepared by standard solid state reaction of ball-milled precursors were investigated as a function of temperature in order to determine the equilibrium phases and establish some kind of a phase diagram in samples with non-stoichiometric compositions [12]. XRD patterns of samples prepared with Fe:Ba = 9 and 7, and sintered at different temperatures, indicated the presence of different equilibrium phases at different temperatures, including BaM, α-Fe_2O_3 (F), $BaFe_2O_4$ spinel (S), and $Ba_3Fe_2O_6$ (B) phases. At a sintering temperature of 800° C, BaM, α-Fe_2O_3, $BaFe_2O_4$ spinel phases appeared as the equilibrium phases, which indicated the formation of the intermediate spinel phase, and the incomplete reaction of this phase with iron oxide to form BaM at such a low temperature. At higher temperatures, the α-Fe_2O_3 phase practically diminished, and the wt. % of BaM increased as a consequence of the completion of the reaction between the spinel and iron oxide phases. The excessive amounts of Ba in the precursor oxide powder were incorporated in the S phase, and B phase which evolve with increasing the sintering temperature. At temperature above 1000° C, the S phase practically disappeared, and all excessive amounts of Ba were incorporated in the B phase which is completed at 1200° C. Although the disappearance of the S phase at temperatures above 1000° C is consistent with previously reported results on solid state chemistry of non-stoichiometric hexaferrites [8], the appearance of the B phase as a high-temperature equilibrium phase was not reported. This phase evolved at temperatures in excess of 900° C, where the spinel phase ceases to be an equilibrium temperature.

The wt. fractions of equilibrium phases at each sintering temperature were determine by Rietveld analysis of the corresponding XRD pattern, and shown in Fig. 19 and Fig. 20. The figures indicated that BaM is the major phase with more than 87% wt. even in the most Fe-deficient sample (Fe;Ba = 7). The wt. % of the B phase (12.43%) and of the BaM phase (87.57%) correspond to a molar ratio of BaM:B = 3.9. The theoretical molar ratio of the two phases as predicted from the initial Fe:Ba ratio of 7 was calculated following the procedure outlined previously [34], and was found to be $(BaM:B)_{th}$ = 3.8. The experimental molar ratio of the two phases is in good agreement with the theoretical value, which confirms the completion of the reaction of the starting powders at 1200° C, resulting in the BaM and B phases as equilibrium phases in the two-component product.

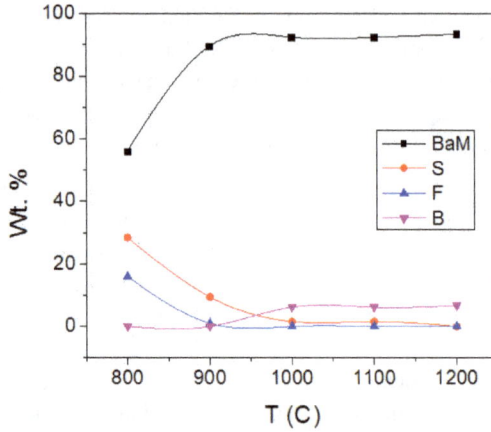

Fig. 19: Relative fractions of the phases present in precursor powders with Fe:Ba = 9.

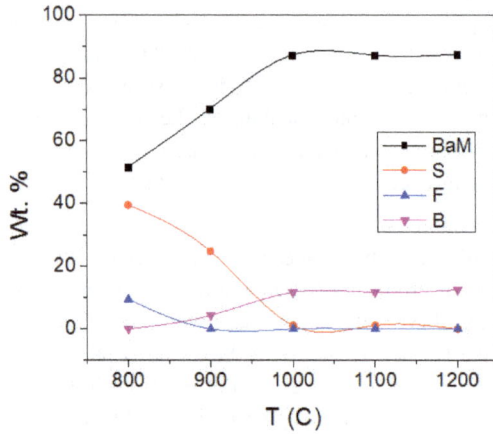

Fig. 20: Relative fractions of the phases present in precursor powders with Fe:Ba =7.

5.2   COPRECIPITATION

This wet chemical technique in which the metal hydroxides are co-precipitated in an aqueous solution of metal salts with the aid of a base (like NaOH) was developed to improve homogeneity of the powder [35], and allow for better reaction of the powder precursors at the molecular level. Solutions of metal chlorides or nitrates with appropriate molar ratios are mixed, and the base is added drop-wise to promote precipitation of metal powders. The co-precipitated powders are washed, dried, and sintered to produce the final BaM product [36]. The sintering temperature required to produce single BaM phase is significantly lower than that required by the conventional solid-state reaction route, and could be as low as 700° C [37]. The starting nitrate or chloride metal solution normally contains Fe:Ba ratios lower the stoichiometric ratio of 12 [38-40]. However, some workers reported better barium ferrite quality at the stoichiometric ratio when sodium carbonate is used as base [40].

BaM powders prepared by this method demonstrated a wide range of magnetic properties, reflecting the sensitivity of the final product to experimental conditions such as stoichiometry, heat treatment, and pH value. In an investigation of the effects of experimental conditions on the properties of BaM prepared by coprecipitation, the particle size was reported to decrease with increasing pH value from 11 to 12.5 for samples prepared from precursor powders with Fe:Ba = 10, and to increase with increasing pH for the samples with Fe:Ba = 10.5 and 11 [41]. In this study, it was found that the sample with Fe:Ba = 10 prepared at pH = 11 and sintered at 920° C exhibited the highest saturation magnetization of 66.1 emu/g, which decreased to 43.6 emu/g at pH = 12.5. The coercivity of this sample, however, increased from 3400 Oe to 4334 Oe with increasing pH, which is consistent with the decrease of the particle size. The sample with Fe:Ba = 11, on the other hand, exhibited the highest coercivity of 4585 Oe at pH = 11, which decreased to 4435 Oe with increasing pH value up to 12.5. The saturation magnetization of this sample decreased from 60.1 emu/g at pH = 11 to 46.2 emu/g at pH = 12.5. In another study, it was demonstrated that increasing the sintering temperature of the coprecipitated powder from 640° C to 920° C improved the saturation magnetization from about 25 emu/g to about 65 emu/g [42]. In this study, the squareness ratio of samples sintered at the different temperatures remained close to 0.5, characteristic of randomly oriented single domain particles, while the coercivity improved slightly from 5264 Oe at 640° C to 5791 Oe at 920° C, which is indicative of single domain particles with high coercivity. On the other hand, BaM prepared from stoichiometric Fe-Ba ratio at pH =12, and sintered at 1300° C, exhibited a saturation magnetization of about 60 emu/g, but a rather low remanence of about 15 emu/g and coercivity of 860 Oe [43]. The low coercivity is probably due to the large particle size (several microns) in this sample.

5.3 SOL-GEL

This method is used to synthesize magnetic powders with controlled particle size distribution. In this method, metal salts (typically nitrates) are dissolved in deionized or distilled water, and the solutions are mixed under constant stirring to ensure the achievement of a homogeneous solution. In the citrate sol-gel method, appropriate molar ratio of citric acid ($C_6H_8O_7$) is then added to the solution under constant stirring. The pH of the solution is adjusted by addition of a basic solution dropwise with constant stirring. The typical value of pH for the solution is around 7 - 9 [37, 44]. The solution is evaporated at about 80° C until a highly viscous gel is formed, and then dried at 200° C. The dried gel is then sintered to produce the required hexaferrite phase.

Experimental conditions including Fe:Ba ratio and heat treatment were found to be key factors in the particle size and morphology, and the magnetic properties of the ferrite powder produced by sol-gel technique. Optimization of the experimental conditions resulted in nanopowders with saturation magnetization of 70 emu/g and coercivity as high as 5950 Oe [45, 46].

5.4 AUTOCOMBUSTION

This is a modified citrate sol-gel method in which self-propagating combustion of the gel occurs upon heating on a hot plate [47-49]. Homogeneous solutions of the metal nitrates and citric acid or ethylene glycol fuel with preset molar ratios are prepared by mixing and stirring. The pH value of the solution is adjusted in the range of 7 – 8 by adding a basic solution [50]. The solution is dehydrated by heating at 80° C on a hot plate. The brown/yellow gel is then heated at 220 – 240° C to induce auto-combustion [51, 52]. Depending on the molar ratio of citric acid to metal nitrates, the wet gel could ignite directly in air, or could only ignite in dried gel form [47, 48]. Auto-combustion yields foamy powder, which is subsequently grinded and sintered at temperatures > 600° C to obtain the final BaM ferrite powder.

The cation-to-fuel ratio was reported to be critical for the quality of the produced ferrites [52]. A ratio of 1:1 resulted in a product with saturation magnetization of 46 emu/g, remanence of 24 emu/g, coercivity of 2587 Oe, and energy product of 0.517 MGOe. The magnetic properties improved significantly upon using a ratio of 1:2, where BaM was produced with saturation magnetization of 55 emu/g, remanence of 28 emu/g, coercivity of 5000 Oe, and energy product of 1.013 MGOe [53]. Also, the Fe:Ba ratio and the calcination temperature were found influential in changing the magnetic properties of the ferrite powder [52]. Ferrite powders calcined at 900° C exhibited a saturation magnetization of about 67 emu/g and a coercivity of 5650 Oe at Fe:Ba = 9, whereas the magnetic properties deteriorated to 51 emu/g and 4700 Oe at Fe:Ba = 11. In addition, the saturation magnetization increased with increasing the calcination temperature from 700° C to 1000° C, and the coercivity increased up to 900° C, and then decreased down to 4500 Oe at 1000° C due to the presence of large particles growing at high temperatures.

5.5   CITRATE PRECURSOR METHOD

This method (also known as the Pecchini method) is used to prepare ultrafine barium ferrite precursor powders at low temperatures via the citrate decomposition route. In this method, stoichiometric solutions of metal salts are mixed with citric acid. The cation-to-citric acid molar ratio is usually 1:1. Ammonia is then added dropwise to the solution to control the pH value and homogenize the solution. The solution is then heated at 80° C to remove excess ammonia, and the resulting clear solution is added drop by drop into ethanol with constant stirring. As a result of alcohol dehydration, this process results in the precipitation of a yellowish barium iron citrate complex, which is subsequently filtered and dried in an oven. The citrate precursor is decomposed at a temperature of 425 – 470° C, and subsequent sintering of the amorphous-like product results in the formation of BaM at temperatures above 600° C [8, 54, 55]. Samples with particle size of 60 nm and saturation magnetization of 61.5 emu/g were obtained at firing temperature of 700° C, while samples fired at 750° C and 800° C exhibited particle growth into the range 80 – 100 nm [56].

5.6   HYDROTHERMAL SYNTHESIS

In this method, water solutions of metal nitrates are transferred into a strong base solution such as NaOH or KOH with the appropriate $OH^-:NO_3^-$ and Fe:Ba ratios. Normally, hydroxide to nitrate ratio is chosen to be 2 [57, 58], and much higher ratios were used by some researchers [59, 60]. Fe:Ba ratios used in this technique are normally lower than stoichiometric ratios [60, 61]. The solution containing the metallic precipitates are then heat treated in an autoclave at temperatures in the range 150 – 290° C. The resulting particles are then filtered, washed and dried in an oven. To improve the magnetic characteristics of the product, the dried powder is sintered at temperatures of 1100 - 1200° C. It was shown that sintering at 1100° C of the hydrothermally synthesized BaM powder with Fe:Ba ratio of 8 leads to an improvement of the saturation magnetization from 46 emu/g to 64 emu/g, and the coercivity peaks at 2.3 kOe at a sintering temperature of 1000° C [61].

5.7   MOLTEN SALT METHOD

This procedure is utilized for the synthesis of large, multi-domain crystals of M-type hexaferrites. In early attempts to produce M-type phase using this technique, the basic procedure involved mixing barium carbonate and iron oxide precursor powders as reactants with NaCl-KCl salt mixture as solvent, and heating the reaction mixture at 800 - 1100° C [62]. The magnetic powder can then be extracted from the dry cake by crushing and washing with distilled water to remove the salts. The quality of the product produced was investigated as a function of experimental conditions such as the starting reactants, solvent composition and purity, heat treat treatment, and other experimental procedures [62, 63]. Large variations of the magnetic properties were found, and hexagonal plates of BaM with basal dimension < 1.5 μm and optimal magnetic properties (saturation magnetization of 72 emu/g, and H_c = 4300 Oe) were synthesized under optimal experimental conditions [62].

In a more sophisticated variation of the method, BaM powder prepared by coprecipitation is used instead. The coprecipitated particles are then mixed with KCl flux at a BaM to salt weight ratio of 1:1. The mixture is initially heated at 450° C, and then at 950° C where particulate BaM phase is produced. The final step is washing the product with deionized water to remove the salts, and then drying at 80° C in an oven [64]. Alternatively, BaM precursor powder was used for the molten salt technique, and the product revealed improvement of the magnetic in comparison with the standard solid state reaction method [65].

5.8   GLASS CRYSTALLIZATION

This method involves mixing the ferrite powder with a glass matrix, and melting the mixture, which is subsequently rapidly quenched to produce an amorphous matrix containing the ferrite. Crystallization of BaM is achieved by annealing at temperatures typically > 600° C. The magnetic ferrite can then be retrieved by dissolving the amorphous matrix in a dilute acid, which does not attack the BaM phase. The heat treatment of the product has a significant influence on the ferrite product, where samples with different heat treatments were reported to possess coercivity ranging from 2600 Oe to 5350 Oe [66].

5.9   SPRAY PYROLYSIS

The system used in the preparation of the ferrite powder consists of an ultrasonic droplet generator, a quartz reactor, and a powder collector [67, 68]. The droplets of the precursor solution are carried to the high temperature reactor by gas flow, which can be adjusted for optimal results. The droplets evaporates and their constituents crystallize at 900° C. The result is a powder consisting of spherical particles, which can be heat treated at 500 - 900° C to improve their magnetic properties. Barium ferrite powders prepared by this method and post-treated at 800° C revealed saturation magnetization close to the theoretical value, and a high coercivity of 6000 Oe [69].

In a variation of the method, BaM ferrite was directly crystallized from an aerosol by using a combination of pyrolysis and citrate precursor methods [70]. Barium iron citrate precursor solution was first prepared with Fe:Ba = 12. Fine droplets of the solution were generated, and passed through a low temperature furnace (200 – 250° C) to evaporate the solvent. A powder composed of submicron hollow spheres was then obtained by passing the dried barium iron citrate particles through a high temperature furnace. The resulting powder had a rather low saturation magnetization of 5.6 emu/g and a coercivity of 2500 Oe. The powder was then heat treated at 1000° C, resulting in a significant improvement of the magnetic properties, where a saturation magnetization of ~ 50.0 emu/g and a coercivity of ~ 5600 Oe were obtained. The heat-treated powder was then hand milled to eliminate the effects of particle aggregation, and this process resulted in an increase of the coercivity up to 5900 Oe, accompanied by a decrease in the saturation magnetization down to 42.6 emu/g. These results indicate that the coercivity of BaM ferrite prepare by spray pyrolysis is probably one of the highest reported for this ferrite.

5.10   ADDITIONAL SYNTHESIS ROUTES

Hexaferrite materials produced by the various synthesis routes discussed above revealed a wide range of magnetic properties, depending on the synthesis method, and the details of adopted experimental conditions. Intensive research work involving the variations of the synthesis routes have been carried out with the objective of improving the quality and magnetic properties of hexaferrite materials. To eliminate unfavorable secondary phases, and enhance the magnetic properties of the final product, various synthesis routes were proposed, including modifications of the experimental conditions, and combining different synthesis routes as demonstrated by our previous discussion. Other modifications of the experimental conditions and synthesis routes were proposed for the production of hexaferrite ceramics, and in the following discussion, we summarize the methodology and results of some research work in this field.

In one of the variations of the synthesis route to produce hexaferrite materials, the oxalate precursor route was adopted for the synthesis of BaM ferrite [71]. Solutions of metal chlorides with Fe:Ba = 12 dissolved in equal amounts of oxalic acid were prepared and mixed with continuous stirring. Dried powder was obtained by heating at 100° C overnight, which was subsequently annealed at different temperatures (800 – 1200° C) for two h to investigate the effect of heat treatment on the properties of the powder. This method generally resulted in the formation of BaM material with low coercivity. The increase of the annealing temperature from 900 to 1100° C, however, was found to result in an increase of both the saturation magnetization and coercivity of the prepared powder. The results of the study revealed a maximum saturation magnetization of 66.36 emu/g, and a maximum coercivity of ~ 640 Oe.

In addition, the ammonium nitrate melt technique (ANMT) [72] was proposed as a new synthesis route for the production of high quality BaM ferrite from starting powders with low Fe:Ba ratios [73]. According to this method, a thick solution was prepared by mixing appropriate amounts of $BaCO_3$ and Fe_2O_3 powders (commensurate with the desired Fe:Ba ratio) with the ammonium nitrate melt and stirring with a magnetic stirrer. The solution was then heated at 260° C to obtain a reddish precipitate, which was subsequently preheated at 450° C for 5 h. The optimal heat treatment was identified by carrying out a systematic study of the properties of parts of the resulting powder subjected to different heat treatments at temperatures in the range of 800 – 1200° C. The results of the study revealed that a high quality BaM ferrite can be produced from mixtures of starting powders with a rather low Fe:Ba ratio. At such low Fe:Ba ratios, however, a mixture of nonmagnetic barium mono-ferrite spinel ($BaFe_2O_4$) and BaM phases was observed in the product, which normally leads to a low saturation magnetization. To overcome this problem, the product powder was washed with a HCl solution, which dissolves the barium spinel phase with no appreciable effect of the BaM phase, leading to the production of the desired pure BaM phase. In particular, the sample prepared from starting powders with Fe:Ba = 2 and sintered at 1100° C gave the best magnetic properties after dissolving away the barium spinel component by HCl washing, where a single BaM phase was obtained with saturation magnetization of 66.7 emu/g, remanent magnetization of 38.5 emu/g, and coercivity of 4228 Oe.

REFERENCES

[1] J. Went, G. Rathenau, E. Gorter, G. Van Oosterhout, Ferroxdure, a class of new permanent magnet materials, Philips Tech. Rev, 13 (1952) 194-208.

[2] J. Went, E. Gorter, The magnetic and electrical properties of ferroxcube materials, Philips Tech. Rev, 13 (1952) 16.

[3] J. Smit, H.P.J. Wijn, Ferrites, Wiley, New York, 1959.

[4] H. Wijn, A New Method of Melting Ferromagnetic Semiconductors. $BaFe_{18}O_{27}$, a New Kind of Ferromagnetic Crystal with High Crystal Anisotropy, (1952).

[5] H.P.J. Wijn, Ferromagnetic domain walls in ferroxdure, Physica, 19 (1953) 555-564.
http://dx.doi.org/10.1016/S0031-8914(53)80061-3

[6]   Ü. Özgür, Y. Alivov, H. Morkoç, Microwave ferrites, part 1: fundamental properties, Journal of Materials Science: Materials in Electronics, 20 (2009) 789-834.
http://dx.doi.org/10.1007/s10854-009-9923-2

[7]   Ü. Özgür, Y. Alivov, H. Morkoç, Microwave ferrites, part 2: passive components and electrical tuning, Journal of Materials Science: Materials in Electronics, 20 (2009) 911-952.
http://dx.doi.org/10.1007/s10854-009-9924-1

[8]   R.C. Pullar, Hexagonal ferrites: a review of the synthesis, properties and applications of hexaferrite ceramics, Progress in Materials Science, 57 (2012) 1191-1334.
http://dx.doi.org/10.1016/j.pmatsci.2012.04.001

[9]   V.G. Harris, A. Geiler, Y. Chen, S.D. Yoon, M. Wu, A. Yang, Z. Chen, P. He, P.V. Parimi, X. Zuo, Recent advances in processing and applications of microwave ferrites, Journal of Magnetism and Magnetic Materials, 321 (2009) 2035-2047.
http://dx.doi.org/10.1016/j.jmmm.2009.01.004

[10]   S.H. Mahmood, M.D. Zaqsaw, O.E. Mohsen, A. Awadallah, I. Bsoul, M. Awawdeh, Q.I. Mohaidat, Modification of the magnetic properties of Co_2Y hexaferrites by divalent and trivalent metal substitutions, Solid State Phenomena, 241 (2016) 93-125.
http://dx.doi.org/10.4028/www.scientific.net/SSP.241.93

[11]   Y. Maswadeh, S.H. Mahmood, A. Awadallah, A.N. Aloqaily, Synthesis and structural characterization of nonstoichiometric barium hexaferrite materials with Fe: Ba ratio of 11.5–16.16, IOP Conference Series: Materials Science and Engineering, IOP Publishing, 2015, pp. 012019.

[12]   Y. Maswadeh, Structural analysis of hexaferrite materials,   Physics, The University of Jordan, 2014.

[13]   X. Obradors, X. Solans, A. Collomb, D. Samaras, J. Rodriguez, M. Pernet, M. Font-Altaba, Crystal structure of strontium hexaferrite $SrFe_{12}O_{19}$, Journal of Solid State Chemistry, 72 (1988) 218-224.
http://dx.doi.org/10.1016/0022-4596(88)90025-4

[14]   E. Gorter, Saturation magnetization of some ferrimagnetic oxides with hexagonal crystal structures, Proceedings of the IEE-Part B: Radio and Electronic Engineering, 104 (1957) 255-260.
http://dx.doi.org/10.1049/pi-b-1.1957.0042

[15] A.M. Alsmadi, I. Bsoul, S.H. Mahmood, G. Alnawashi, K. Prokeš, K. Siemensmeyer, B. Klemke, H. Nakotte, Magnetic study of M-type doped barium hexaferrite nanocrystalline particles, Journal of Applied Physics, 114 (2013) 243910.
http://dx.doi.org/10.1063/1.4858383

[16] B.D. Cullity, C.D. Graham, Introduction to magnetic materials, John Wiley & Sons2011.

[17] G.H. Dushaq, S.H. Mahmood, I. Bsoul, H.K. Juwhari, B. Lahlouh, M.A. AlDamen, Effects of molybdenum concentration and valence state on the structural and magnetic properties of $BaFe_{11.6}Mo_xZn_{0.4-x}O_{19}$ hexaferrites, Acta Metallurgica Sinica (English Letters), 26 (2013) 509-516.
http://dx.doi.org/10.1007/s40195-013-0075-2

[18] S.H. Mahmood, G.H. Dushaq, I. Bsoul, M. Awawdeh, H.K. Juwhari, B.I. Lahlouh, M.A. AlDamen, Magnetic Properties and Hyperfine Interactions in M-Type $BaFe_{12-2x}Mo_xZn_xO_{19}$ Hexaferrites, Journal of Applied Mathematics and Physics, 2 (2014) 77-87.
http://dx.doi.org/10.4236/jamp.2014.25011

[19] S. Mahmood, A. Aloqaily, Y. Maswadeh, A. Awadallah, I. Bsoul, H. Juwhari, Structural and Magnetic Properties of Mo-Zn Substituted $(BaFe_{12-4x}Mo_xZn_{3x}O_{19})$ M-Type Hexaferrites, Material Science Research India, 11 (2014) 09-20.

[20] M. Awawdeh, I. Bsoul, S.H. Mahmood, Magnetic properties and Mössbauer spectroscopy on Ga, Al, and Cr substituted hexaferrites, Journal of Alloys and Compounds, 585 (2014) 465-473.
http://dx.doi.org/10.1016/j.jallcom.2013.09.174

[21] G. Bate, Recording materials, in: P. E, Wohlfarth (Ed.) Ferromagnetic materials, North-Holland Publishing Company, New York, 1980, pp. 381-508.

[22] Joonghoe Dho, E.K. Lee, N.H.H. J.Y. Park, Effects of the grain boundary on the coercivity of barium ferrite $BaFe_{12}O_{19}$, Journal of Magnetism and Magnetic Materials, 285 (2005) 164-168.
http://dx.doi.org/10.1016/j.jmmm.2004.07.033

[23] A. Stuijts, G. Rathenau, G. Weber, Ferroxdure ii and iii, anisotropic permanent magnet materials, Philips Tech. Rev, 16 (1954) 7.

[24] I.Y. Gershov, Barium ferrite permanent magnets, Soviet Powder Metallurgy and Metal Ceramics, 1 (1964) 386-393.
http://dx.doi.org/10.1007/BF00774124

[25] H. Fu, H.R. Zhai, H.C. Zhang, B.X. Gu, J.Y. Li, Magnetic properties on Mn substituted barium ferrite, Journal of Magnetism and Magnetic Materials, 54-57 (1986) 905-906.
http://dx.doi.org/10.1016/0304-8853(86)90307-0

[26] I. Bsoul, S.H. Mahmood, A.F. Lehlooh, Structural and magnetic properties of $BaFe_{12-2x}Ti_xRu_xO_{19}$, Journal of Alloys and Compounds, 498 (2010) 157-161.
http://dx.doi.org/10.1016/j.jallcom.2010.03.142

[27] I. Bsoul, S.H. Mahmood, Magnetic and structural properties of $BaFe_{12-x}Ga_xO_{19}$ nanoparticles, Journal of Alloys and Compounds, 489 (2010) 110-114.
http://dx.doi.org/10.1016/j.jallcom.2009.09.024

[28] A. Alsmadi, I. Bsoul, S. Mahmood, G. Alnawashi, F. Al-Dweri, Y. Maswadeh, U. Welp, Magnetic study of M-type Ru-Ti doped strontium hexaferrite nanocrystalline particles, Journal of Alloys and Compounds, 648 (2015) 419-427.
http://dx.doi.org/10.1016/j.jallcom.2015.06.274

[29] I. Bsoul, S. Mahmood, Structural and magnetic properties of $BaFe_{12-x}Al_xO_{19}$ prepared by milling and calcination, Jordan J. Phys., 2 (2009) 171-179.

[30] G. Turilli, F. Licci, S. Rinaldi, A. Deriu, Mn^{2+}, Ti^{4+} substituted barium ferrite, Journal of Magnetism and Magnetic Materials, 59 (1986) 127-131.
http://dx.doi.org/10.1016/0304-8853(86)90019-3

[31] A. Awadallah, S.H. Mahmood, Y. Maswadeh, I. Bsoul, M. Awawdeh, Q.I. Mohaidat, H. Juwhari, Structural, magnetic, and Mossbauer spectroscopy of Cu substituted M-type hexaferrites, Materials Research Bulletin, 74 (2016) 192-201.
http://dx.doi.org/10.1016/j.materresbull.2015.10.034

[32] O.T. Ozkan, H. Erkalfa, The effect of B_2O_3 addition on the direct sintering of barium hexaferrite, Journal of the European Ceramic Society, 14 (1994) 351-358.
http://dx.doi.org/10.1016/0955-2219(94)90072-8

[33] P. Hernandez-Gomez, J.M. Munoz, C. Torres, C. de Francisco, O. Alejos, Influence of stoichiometry on the magnetic disaccommodation in barium M-type hexaferrites, Journal of Physics D: Applied Physics, 36 (2003) 1062-1070.
http://dx.doi.org/10.1088/0022-3727/36/9/303

[34] S.H. Mahmood, A.N. Aloqaily, Y. Maswadeh, A. Awadallah, I. Bsoul, M. Awawdeh, H.K. Juwhari, Effects of heat treatment on the phase evolution, structural, and magnetic properties of Mo-Zn doped M-type hexaferrites, Solid State Phenomena, 232 (2015) 65-92.
http://dx.doi.org/10.4028/www.scientific.net/SSP.232.65

[35]   P. Garcia-Casillas, A. Beesley, D. Bueno, J. Matutes-Aquino, C. Martinez, Remanence properties of barium hexaferrite, Journal of alloys and compounds, 369 (2004) 185-189.
http://dx.doi.org/10.1016/j.jallcom.2003.09.100

[36]   D. Lisjak, M. Drofenik, The mechanism of the low-temperature formation of barium hexaferrite, Journal of the European Ceramic Society, 27 (2007) 4515-4520.
http://dx.doi.org/10.1016/j.jeurceramsoc.2007.02.202

[37]   J.-P. Wang, L. Ying, M.-L. Zhang, Y.-j. QIAO, X. Tian, Comparison of the Sol-gel Method with the Coprecipitation Technique for Preparation of Hexagonal Barium Ferrite, Chemical Research in Chinese Universities, 24 (2008) 525-528.
http://dx.doi.org/10.1016/S1005-9040(08)60110-5

[38]   H.B. von Basel, K.A. Hempel, Static magnetic properties of pressure-sintered barium ferrite, Journal of Magnetism and Magnetic Materials, 38 (1983) 316-318.
http://dx.doi.org/10.1016/0304-8853(83)90373-6

[39]   S.E. Jacobo, C. Domingo-Pascual, R. Rodrigez-Clemente, M.A. Blesa, Synthesis of ultrafine particles of barium ferrite by chemical coprecipitation, Journal of Materials Science, 33 (1997) 1025-1028.
http://dx.doi.org/10.1023/A:1018582423406

[40]   M. Rashad, I. Ibrahim, A novel approach for synthesis of M-type hexaferrites nanopowders via the co-precipitation method, Journal of Materials Science: Materials in Electronics, 22 (2011) 1796-1803.
http://dx.doi.org/10.1007/s10854-011-0365-2

[41]   S.R. Janasi, D. Rodrigues, F.J. Landgraf, M. Emura, Magnetic properties of coprecipitated barium ferrite powders as a function of synthesis conditions, Magnetics, IEEE Transactions on, 36 (2000) 3327-3329.
http://dx.doi.org/10.1109/intmag.2000.872434

[42]   J. Matutes-Aquino, S. Dıaz-Castanón, M. Mirabal-Garcıa, S. Palomares-Sánchez, Synthesis by coprecipitation and study of barium hexaferrite powders, Scripta materialia, 42 (2000) 295-299.
http://dx.doi.org/10.1016/S1359-6462(99)00350-4

[43]   P. Shepherd, K.K. Mallick, R.J. Green, Magnetic and structural properties of M-type barium hexaferrite prepared by co-precipitation, Journal of magnetism and magnetic materials, 311 (2007) 683-692.
http://dx.doi.org/10.1016/j.jmmm.2006.08.046

[44] Z. Mosleh, P. Kameli, A. Poorbaferani, M. Ranjbar, H. Salamati, Structural, magnetic and microwave absorption properties of Ce-doped barium hexaferrite, Journal of Magnetism and Magnetic Materials, 397 (2016) 101-107. *http://dx.doi.org/10.1016/j.jmmm.2015.08.078*

[45] C. Sürig, D. Bonnenberg, K. Hempel, P. Karduck, H. Klaar, C. Sauer, Effects of Variations in Stoichiometry on M-Type Hexaferrites, Le Journal de Physique IV, 7 (1997) C1-315-C311-316.

[46] W. Zhong, W. Ding, N. Zhang, J. Hong, Q. Yan, Y. Du, Key step in synthesis of ultrafine $BaFe_{12}O_{19}$ by sol-gel technique, Journal of Magnetism and Magnetic Materials, 168 (1997) 196-202. *http://dx.doi.org/10.1016/S0304-8853(96)00664-6*

[47] R.C. Alange, P.P. Khirade, S.D. Birajdar, A.V. Humbe, K.M. Jadhav, Structural, magnetic and dielectric properties of Al-Cr co-substituted M-type barium hexaferrite nanoparticles, Journal of Molecular Structure, 1106 (2016) 460-467. *http://dx.doi.org/10.1016/j.molstruc.2015.11.004*

[48] Y. Hong, C. Ho, H.Y. Hsu, C. Liu, Synthesis of nanocrystalline $Ba(MnTi)_xFe_{12-2x}O_{19}$ powders by the sol–gel combustion method in citrate acid–metal nitrates system (x = 0, 0.5, 1.0, 1.5, 2.0), Journal of magnetism and magnetic materials, 279 (2004) 401-410. *http://dx.doi.org/10.1016/j.jmmm.2004.02.008*

[49] S.H. Mahmood, F.S. Jaradat, A.F. Lehlooh, A. Hammoudeh, Structural properties and hyperfine interactions in Co-Zn Y-type hexaferrites prepared by sol-gel method, Ceramics International, 40 (2014) 5231-5236. *http://dx.doi.org/10.1016/j.ceramint.2013.10.092*

[50] W. Abbas, I. Ahmad, M. Kanwal, G. Murtaza, I. Ali, M.A. Khan, M.N. Akhtar, M. Ahmad, Structural and magnetic behavior of Pr-substituted M-type hexagonal ferrites synthesized by sol–gel autocombustion for a variety of applications, Journal of Magnetism and Magnetic Materials, 374 (2015) 187-191. *http://dx.doi.org/10.1016/j.jmmm.2014.08.029*

[51] Simon Thompson, Neil J. Shirtcliffe, Eoin S. O'Keefe, Steve Appleton, C.C. Perry, Synthesis of $SrCo_xTi^xFe_{(12-2x)}O_{19}$ through sol-gel auto-ignition and its characterisation, Journal of Magnetism and Magnetic Materials, 297 (2005) 100-1007. *http://dx.doi.org/10.1016/j.jmmm.2004.10.102*

[52]   Y. Meng, M. He, Q. Zeng, D. Jiao, S. Shukla, R. Ramanujan, Z. Liu, Synthesis of barium ferrite ultrafine powders by a sol–gel combustion method using glycine gels, Journal of Alloys and Compounds, 583 (2014) 220-225.
http://dx.doi.org/10.1016/j.jallcom.2013.08.156

[53]   D. Bahadur, S. Rajakumar, A. Kumar, Influence of fuel ratios on auto combustion synthesis of barium ferrite nano particles, Journal of chemical sciences, 118 (2006) 15-21.
http://dx.doi.org/10.1007/BF02708761

[54]   V. Sankaranarayanan, Q. Pankhurst, D. Dickson, C. Johnson, Ultrafine particles of barium ferrite from a citrate precursor, Journal of magnetism and magnetic materials, 120 (1993) 73-75.
http://dx.doi.org/10.1016/0304-8853(93)91290-N

[55]   V. Sankaranarayanan, D. Khan, Mechanism of the formation of nanoscale M-type barium hexaferrite in the citrate precursor method, Journal of magnetism and magnetic materials, 153 (1996) 337-346.
http://dx.doi.org/10.1016/0304-8853(95)00537-4

[56]   V. Sankaranarayanan, Q. Pankhurst, D. Dickson, C. Johnson, An investigation of particle size effects in ultrafine barium ferrite, Journal of magnetism and magnetic materials, 125 (1993) 199-208.
http://dx.doi.org/10.1016/0304-8853(93)90838-S

[57]   A. Ataie, I. Harris, C. Ponton, Magnetic properties of hydrothermally synthesized strontium hexaferrite as a function of synthesis conditions, Journal of materials science, 30 (1995) 1429-1433.
http://dx.doi.org/10.1007/BF00375243

[58]   L. Lechevallier, J. Le Breton, J. Wang, I. Harris, Structural analysis of hydrothermally synthesized $Sr_{1-x}Sm_xFe_{12}O_{19}$ hexagonal ferrites, Journal of magnetism and magnetic materials, 269 (2004) 192-196.
http://dx.doi.org/10.1016/S0304-8853(03)00591-2

[59]   D. Primc, D. Makovec, D. Lisjak, M. Drofenik, Hydrothermal synthesis of ultrafine barium hexaferrite nanoparticles and the preparation of their stable suspensions, Nanotechnology, 20 (2009) 315605.
http://dx.doi.org/10.1088/0957-4484/20/31/315605

[60]   M. Drofenik, I. Ban, D. Makovec, A. Žnidaršič, Z. Jagličić, D. Hanžel, D. Lisjak, The hydrothermal synthesis of super-paramagnetic barium hexaferrite particles, Materials Chemistry and Physics, 127 (2011) 415-419.
http://dx.doi.org/10.1016/j.matchemphys.2011.02.037

[61]   X. Liu, J. Wang, L.-M. Gan, S.-C. Ng, Improving the magnetic properties of hydrothermally synthesized barium ferrite, Journal of magnetism and magnetic materials, 195 (1999) 452-459.
http://dx.doi.org/10.1016/S0304-8853(99)00123-7

[62]   R.H. Arendt, The molten salt synthesis of single domain $BaFe_{12}O_{19}$ and $SrFe_{12}O_{19}$ crystals, Journal of Solid State Chemistry, 8 (1973) 339-347.
http://dx.doi.org/10.1016/S0022-4596(73)80031-3

[63]   T.-S. Chin, S. Hsu, M. Deng, Barium ferrite particulates prepared by a salt-melt method, Journal of magnetism and magnetic materials, 120 (1993) 64-68.
http://dx.doi.org/10.1016/0304-8853(93)91288-I

[64]   Y. Liu, M.G. Drew, Y. Liu, J. Wang, M. Zhang, Preparation, characterization and magnetic properties of the doped barium hexaferrites $BaFe_{12-2x}Co_{x/2}Zn_{x/2}Sn_xO_{19}$, x = 0.0–2.0, Journal of Magnetism and Magnetic Materials, 322 (2010) 814-818.
http://dx.doi.org/10.1016/j.jmmm.2009.11.009

[65]   S. Dursun, R. Topkaya, N. Akdoğan, S. Alkoy, Comparison of the structural and magnetic properties of submicron barium hexaferrite powders prepared by molten salt and solid state calcination routes, Ceramics International, 38 (2012) 3801-3806.
http://dx.doi.org/10.1016/j.ceramint.2012.01.028

[66]   B. Shirk, W. Buessem, Magnetic properties of barium ferrite formed by crystallization of a glass, Journal of the American Ceramic Society, 53 (1970) 192-196.
http://dx.doi.org/10.1111/j.1151-2916.1970.tb12069.x

[67]   D. Jung, S. Hong, J. Cho, Y. Kang, Nano-sized barium titanate powders with tetragonal crystal structure prepared by flame spray pyrolysis, Journal of the European Ceramic Society, 28 (2008) 109-115.
http://dx.doi.org/10.1016/j.jeurceramsoc.2007.05.018

[68]   J.S. Cho, D.S. Jung, S.K. Hong, Y.C. Kang, Characteristics of nano-sized pb-based glass powders by high temperature spray pyrolysis method, Journal of the Ceramic Society of Japan, 116 (2008) 600-604.
http://dx.doi.org/10.2109/jcersj2.116.600

[69]   M.H. Kim, D.S. Jung, Y.C. Kang, J.H. Choi, Nanosized barium ferrite powders prepared by spray pyrolysis from citric acid solution, Ceramics International, 35 (2009) 1933-1937.
http://dx.doi.org/10.1016/j.ceramint.2008.10.016

[70]   T. Gonzalez-Carreno, M. Morales, C. Serna, Barium ferrite nanoparticles prepared directly by aerosol pyrolysis, materials letters, 43 (2000) 97-101.

[71]   Q. Mohsen, Barium hexaferrite synthesis by oxalate precursor route, Journal of Alloys and Compounds, 500 (2010) 125-128.
       http://dx.doi.org/10.1016/j.jallcom.2010.03.230

[72]   U. Topal, H. Ozkan, H. Sozeri, Synthesis and characterization of nanocrystalline $BaFe_{12}O_{19}$ obtained at 850 C by using ammonium nitrate melt, Journal of magnetism and magnetic materials, 284 (2004) 416-422.
       http://dx.doi.org/10.1016/j.jmmm.2004.07.009

[73]   U. Topal, H. Ozkan, L. Dorosinskii, Finding optimal Fe/Ba ratio to obtain single phase $BaFe_{12}O_{19}$ prepared by ammonium nitrate melt technique, Journal of alloys and compounds, 428 (2007) 17-21.
       http://dx.doi.org/10.1016/j.jallcom.2006.03.047

CHAPTER 4

Ferrites with High Magnetic Parameters

S.H. Mahmood

Physics Department, The University of Jordan, Amman, Jordan

s.mahmood@ju.edu.jo

Abstract

Hexaferrites with chemical formula $MFe_{12}O_{19}$ (M = Ba, Sr, Pb) exhibit properties suitable for a wide range of applications. The magnetic properties of these ferrites require tuning to fit the requirements of a given application. This can be achieved by modifying the experimental techniques, as well as adopting strategic metal substitutions in the ferrite lattice. The effects of experimental techniques and conditions on the magnetic properties of M-type hexaferrites are reviewed. Also, the effects of various scenarios of metal substitutions on the improvement of the magnetic properties are discussed.

Keywords

M-Type Hexaferrite; Structural Properties; Magnetic Properties; Synthesis

Contents

1. INTRODUCTION

The demand for higher performance magnets had promoted vigorous scientific research concerned with the development of new methods, and adopting various scenarios with the aim of improving the magnetic properties of permanent magnet materials in use. This was reflected by the exponential growth of the number of published research articles on hexaferrites since their inception in the early 1950s (Fig. 1 in the previous chapter). The desired modification of the magnetic properties of a material is, however, highly dependent on the type of application it is designed for. Permanently magnetizable material with the highest possible remanent flux density and coercivity, for example, is required for the production of permanent magnets to be used in high power motors and drives. Theoretically, such a material would be ideal for magnetic data storage media applications due to its high resistivity to demagnetization by stray fields, and the consequent loss of recorded information. The magnetic hardness of the material, however, requires high magnetizing fields for data recording, which is translated as higher power consumption in the recording process, and large magnetizing magnets. In principle, such a hard material can be used for special magnetic recording applications, when information is desired to be permanently recorded, and the storage medium is not required to be reused. Practically, the demand for device miniaturization, in addition to the requirements of the reduction of power consumption, necessitates the compromise for lower coercivity in magnetic recording applications. Further, tuning the magnetocrystalline anisotropy of the ferrite is crucial for the design of microwave devices operating in a specific frequency range [1-4].

The requirements for industrial and technological applications cannot always be met by the utilization of standard M-type hexaferrites, and call for the development of hexaferrite materials with properties tuned for the desired application. To this end, a great wealth of literature is available as a consequence of the continuous human efforts to improve the magnetic properties of hexaferrite materials for different practical applications. Such efforts spanned the widest space available for materials development, including modifications of the fabrication methods, variations of the experimental conditions, and variations of the cationic substitutions and chemical stoichiometries. In some cases, complex procedures involving modified methods of synthesis, varied experimental conditions, and changes in chemical stoichiometry are adopted. Therefore, it would be difficult to cover all aspects of hexaferrite materials development, and classify the effects of the different types of modifications adopted into separate categories with sharp boundaries. The following sections are organized to address some results of the most important research activities toward hexaferrite materials development and characterization.

At this point, it is worth mentioning that different authors focused on different magnetic properties, and sometimes used different units for the magnetic parameters. Also, the quoted saturation magnetization by some authors was merely the magnetization at the maximum applied field, which was in some cases as low as 8000 Oe, significantly lower than the field required for true saturation of the sample, which is normally > 12000 Oe. In order to make clear comparison between the results of various investigators, we converted the quoted results into cgs system of units (emu/g for magnetization, and Oe and G for magnetic field and induction). Sometimes, the figures throughout our discussion were rounded up for brevity and significance.

2. EFFECTS OF SYNTHESIS AND EXPERIMENTAL CONDITIONS

In the previous chapter, the major physical and chemical methods utilized for the synthesis of barium hexaferrite precursor powders were briefly discussed. Also, some aspects of the effects of the experimental conditions on the magnetic properties of the ferrite powders produced by each method were addressed. In general, chemical methods are used to improve reaction leading to the production of a more homogeneous ferrite powder with controlled particle size distribution.

The particle size and the proportions of secondary phases in the final product are key factors in determining the quality of the synthesized magnetic powder. Specifically, non-magnetic impurity phases lead to an undesirable reduction of the saturation magnetization, although the coercivity could not be influenced by the presence of such phases. As a consequence, the maximum energy product of the fabricated ferrite magnet is reduced. On the other hand, the optimal particle size for the highest magnetic property is just below the critical single domain size, which is typically 0.5 – 1 µm [1]. Powders with much less particle size usually demonstrate low coercivity, which can be quenched in powders exhibiting superparamagnetic behavior due to their small particle size. On the other hand, the presence of large particles in the powder would also reduce the coercivity as a consequence of domain wall motion during the magnetization-demagnetization processes. Therefore, the control of the particle size in a highly pure ferrite powder is critical for the production of high quality magnetic materials for high performance magnets. Accordingly, intensive research activities were carried out for these purposes, which ultimately lead to the development of variable and complex experimental procedures. Among these, chemical methods demonstrated success in synthesizing powders with coercivity close to the theoretical values. However, one must not ignore the effects of experimental conditions adopted in a given synthesis route, which may lead to deterioration of the magnetic properties, regardless of the superiority of the technique over other techniques.

Among the various synthesis routes of hexaferrite powders discussed in the previous chapter, the conventional ceramic method, in addition to chemical coprecipitation and sol–gel methods, received the greatest interest of researchers. In the following subsections, we briefly review some of the results of research work concerning the effects of experimental conditions adopted in synthesizing BaM ferrite by these three experimental techniques.

2.1. CONVENTIONAL CERAMIC METHOD

Conventional ceramic method based on mixing appropriate proportions of α-Fe_2O_3 with $BaCO_3$, finely grinding, and sintering at temperatures $\gtrsim 1100°$ C is a common method for the production of industrial grade ferrite magnets. The precursor ferrite powder is compacted at high pressure into the desired shape, and sintered for near full densification ($\sim$90% of the theoretical density of 5.28 g/cm^3) [5]. In the early stages of the synthesis of BaM ferrites by the conventional ceramic method, stoichiometric ratios of metal oxides were mixed homogeneously and fired at temperatures above 1000° C [6]. The reported saturation magnetization of a polycrystalline sample at liquid helium temperature was $\sim$ 100 emu/g, corresponding to 20 μ_B per molecule, which is consistent with a collinear magnetic structure in $BaFe_{12}O_{19}$ hexaferrite. Room temperature saturation magnetization and Curie temperature of this compound were 72 emu/g and 450° C, respectively [7].

The effects of different experimental conditions on the magnetic properties of BaM ferrite powders prepared by the conventional ceramic method was of concern to scientists since the early development stages of this ferrite [7-9]. At sintering temperatures below 1000° C, barium mono-ferrite ($BaFe_2O_4$) and α-Fe_2O_3 intermediate phases are normally observed, while complete reaction of these phases to form BaM from stoichiometric starting powders require higher temperatures. Such high sintering temperatures are undesirable for certain devices such as multi-layer chip inductors, where a conductive layer of melting point below 1000° C (such as silver) is used to separate the ferrite layers. In such cases, lowering the formation temperature of the M-type phase is crucial.

The formation temperature of BaM ferrite prepared by the conventional solid-state reaction was lowered by the addition of small amounts of $BaCu(B_2O_5)$, and the density of BaM added only with 1 wt.% $BaCu(B_2O_5)$ was reported to increase from $\sim$ 75% up to $\sim$ 90% with increasing the sintering temperature from 850° C to 950° C [10]. Further increase of the $BaCu(B_2O_5)$ weight fraction to 3% improved the magnet density up to 92%, and resulted in an enhancement of the saturation magnetization from 55.6 emu/g up to 61.4 emu/g. Also, small additions of B_2O_3 was found to result in a single BaM phase with enhanced remanent magnetization at 900° C [11]. The remanent magnetization, saturation magnetization, and coercivity of the ferrite added with 1 wt.% B_2O_3 were found to be 28 emu/g, 54 emu/g, and 2000 – 3000 Oe, respectively. The relatively high remanent magnetization and moderate coercivity of the product are in the range of properties required for high density magnetic recording applications. The magnetic properties of B_2O_3- added BaM ferrite were further improved by etching the powder with diluted HCl solution to remove residual nonmagnetic phases [12]. This procedure applied to the sample added with 0.1 wt. % B_2O_3 resulted in an improvement of the remanent magnetization up to 34.9 emu/g), and saturation magnetization up to 63.3 emu/g, without any systematic influence on the coercivity.

Investigation of the effect of Fe:Ba molar ratio and heat treatment on the equilibrium phases and magnetic properties of BaM ferrites was the subject of interest for several researchers [8, 13-15]. Normally, Fe:Ba ratios somewhat lower than the stoichiometric value of 12 are used to improve the properties of the product [8]. BaM magnetic powders prepared from mixtures with Fe:Ba $\geq$ 12 normally contain a fraction of unreacted α-Fe_2O_3 phase, while starting powder mixtures with Fe:Ba < 11 developed $BaFe_2O_4$ spinel as a secondary phase coexisting with the major M-type phase [16]. From magnetic data, early investigations revealed that the optimum Fe:Ba ratio is about 11.0 [8]. The influence of increased sintering temperature is manifested by the particle growth, which may induce significant reduction of the coercivity, and possibly slight improvement of the saturation magnetization. BaM powders prepared at sintering temperature of 1350° C and milled for different times (from 10 min to 5400 min) to reduce the particle size were investigated [17]. The sample with particle size of 7.1 μm exhibited a coercivity of 725 Oe, which increased steadily up to 2602 Oe for the sample with particle size 1.4 μm. In all these samples, the role of domain wall motion in reducing the coercivity was evident, which is consistent with mean particle size larger than the critical single domain size. A significant improvement of the coercivity was achieved by extended milling of BaM powder and annealing at 1000° C, where a coercivity up to 5653 Oe was obtained, with a saturation magnetization ~ 90% of the theoretical value [18]. Also, the variation of Fe:Ba ratio between 11 and 12 was found to induce a small change in the magnetic properties of the hexaferrite, and the optimal ratio of 11.8 was reported to give BaM with a maximum magnetization (at 10 kOe) of 52.3 emu/g, a remanent magnetization of 34.2 emu/g, and a coercivity of 4537 Oe [19].

From the above discussion, it is clear that only little can be done to improve the magnetic properties of BaM ferrite by variation of the experimental conditions in the conventional ceramic route. Other chemical based methods, such as coprecipitation and sol-gel techniques, proved more efficient in improving the magnetic properties for permanent magnet applications. Specifically more than 50% improvement of the coercivity was achieved by these methods under controlled experimental conditions. In the forthcoming sections we review some of the main results achieved by these methods.

2.2. COPRECIPITATION

Coprecipitation technique is widely used to prepare BaM powder with narrow grain size distributin and high coercivity at relativly low sintering temperatures [20-23]. Experimental conditions such as Fe:Ba ratio, heat treatment, and pH are of significat importance for the quality and characteristics of the magnetic powder product. Accordingly, considerable research work was devoted to the investigation of the effects of different experimental conditions on the properties of BaM ferrites prepared by the coprecipitation method [21].

Variations among the experimental conditions are critical in controlling the grain size and morphology of the prepared ferrite powder, and modiying its magnetic properties. A systematic study was carried out to investigate the effects of the different combinations of experimental parameters (Ba:Fe = 10, 10.5, and 11; pH = 11, 12, and 12.5; sintering temperature = 860° C, 920° C, and 1000° C) on the quality and properties of BaM ferrite powders prepared by coprecipitation [21]. In this study, the grain size of BaM hexaferrite with Fe:Ba = 10 was reported to decrease with increasing pH from 11 to 12.5, while for the samples with Fe:Ba = 10.5 or 11, the grain size increased with increasing pH. The highest saturation magnetization of 66.1 emu/g was obtained for the sample with Fe:Ba = 10 prepared under pH = 11 and sintered at 920° C. The increase in pH during the coprecipitation resulted in a systematic decrease in saturation magnetization, reaching 43.6 emu/g at pH = 12.5. On the other hand, increasing the pH resulted in a systematic increase of the coercivity from 3400 Oe to 4334 Oe, which is consistent with the decrease of the particle size. Powders prepared with Fe:Ba = 11 at sintering temperature of 920° C, however, exhibited the highest coercivity of 4585 Oe at pH = 11, which slightly decreased down to 4435 Oe with increasing pH value up to 12.5. The behavior of the saturation magnetization of this sample with increasing pH, however, is similar to that of the sample with Fe:Ba = 10, but decreasing from 60.1 emu/g at pH = 11 down to 46.2 emu/g at pH = 12.5. In addition, BaM precursor powders with Fe:Ba = 10.5 exhibited lower saturation magnetization (36.5 emu/g – 49.7 emu/g), and different behaviors of the saturation magnetization and coercivity.

The effect of the variation of the heating strategies at otherwise fixed experimental conditions on the magnetic properties of ferrites prepared by coprecipitation also attracted the attention of researchers. BaM ferrites prepared from metal chlorides with Fe:Ba = 10.1 at pH = 12 – 13 revealed almost similar coercivities of powders heated for 4 h at temperatures between 950° C and 1100° C, with the highest coercivity of 5291 Oe observed at 1000° C [24]. At higher temperatures, the coercivity decreased, reaching 3965 Oe at 1200° C. The saturation magnetization, on the other hand, improved significantly from 57.5 emu/g at 950° C to slightly above 75 emu/g at temperatures in the range of 1000 – 1100° C, and the remanent magnetization improved from 30 emu/g to about 39 emu/g. The effect of heat treatment on the properties of BaM prepared by coprecipitation using metal nitrate solutions with Fe:Ba = 11 at pH = 13 was also investigated in the temperature range from 600° C to 1100° C [25]. At temperatures between 700° C and 1100° C, the saturation magnetization fluctuated between ~ 50 – 55 emu/g with no systematic behavior. The coercivity, however, increased up to the range ~ 5300 Oe – 5700 Oe at calcination temperatures between 700° C and 800° C, and then decreased almost linearly with further increasing the temperature, reaching ~ 2100 Oe at 1100° C. The major drop in coercivity to below 4000 Oe was observed in samples with mean particle diameters > 0.2 μm. However, an appreciably higher coercivity of 4639 was reported for a sample composed of 0.6 μm particles [26].

The formation temperature of BaM phase prepared by chemical coprecipitation was found to be dependent on the crystallization mechanisms from the coprecipitated precursors. Complete crystallization of BaM phase prepared from chloride solutions was achieved at 700° C, and the saturation magnetization increased slightly from 68.5 emu/g to 70 emu/g, while the coercivity dropped slightly from 5078 Oe to 5044 Oe as the temperature was raised to 950° C [22]. The results of this study revealed that the direct crystallization of BaM phase from an amorphous precursor resulted in the lowering of the M-type formation temperature, which is consistent with the results of others [27-29]. However, the type of the starting solutions (chloride versus nitrate) and solvent (water versus ethanol), as well as the calcination temperature and soaking time, were reported to play a critical role in the magnetic properties of the ferrite powder [28]. Also, an increase of BaM heating temperature from 750° C to 920° C resulted in a small increase in saturation, remanent magnetization, and coercivity from 63.71 emu/g, 34.88 emu/g, and 5691 Oe to 64.78 emu/g, 35.66 emu/g, and 5791 Oe, respectively [30].

The coercivity of the ferrite powder, however, could be critically influenced by the particle size, which is highly dependent on the experimental conditions and the synthesis route. A significant decrease in coercivity is observed for ferrites prepared under experimental conditions leading to the formation of multi-domain particle assembly [31]. Specifically, BaM samples prepared with stoichiometric Fe:Ba ratio at pH = 12, and sintered at 1300° C, exhibited grain size ranging from 3.5 to > 10 μm with saturation magnetization of ~ 60 emu/g and a rather low coercivity of 860 Oe [32]. Taking into consideration that the sample consisted of a pure BaM phase, the observed low coercivity is obviously an indication of domain wall dominated magnetization mechanism in the large multi domain particles. On the other hand, a rather low coercivity of 221.8 Oe was reported for BaM phase composed of particles with diameters in the range of 49 – 82 nm, with a relatively high saturation magnetization of 60.8 emu/g and remanent magnetization of 31.0 emu/g [27]. The reported squareness ratio of 0.51 was very close to the theoretical value of 0.50 for an assembly of randomly oriented single domain particles, and the low coercivity of the sample was associated with the superparamagnetic behavior of the ultrafine particles.

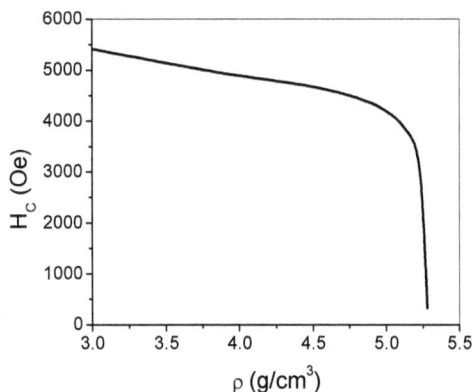

Fig. 1. The intrinsic coercivity of BaM ferrites vs. density (reproduced from the data in Ref. [33])

A systematic study of pressure-sintered BaM ferrites prepared by coprecipitation indicated that the coercivity of the material could be correlated with the density of the magnet [33]. The coercivty was reported to decrease slowly from about 5400 Oe for a 57% dense magnet (ρ = 3.0 g/cm^3) to about 4300 Oe for a 95% dense magnet (ρ = 5.0 g/cm^3). At higher densities, the coecivity dropped sharply, approacing that characteristic of a soft magnet at near theoretical density. This behavior, illustrated by Fig. 1, was correlated with different grain growth mechanisms. Generally, the study revealed that the density of the ferrite prepared with a given Fe:Ba ratio (between 10.6 and 11.8) increased with increasing the sintering temperature from 1050° C to 1150° C, and with increasing the presure-sintering time from 2 – 60 min. Further, in an earlier study, the effects of sintering temperature and magnet densification on the magnetic properties of BaM powders prepared by coprecipitation with a narrow size distribution (20 – 150 nm) were investigated [23]. This study revealed the best magnetic properties for samples prepared by pelletizing under pressure of 245 MPa and sintering at 1100° C: ρ = 4.6 g/cm^3, H_c = 5.45 kOe, remanent magnetization = 35.6 emu/g, and $(BH)_{max}$ = 0.93 MGOe. It was also found that sintering at higher temperatures deteriorates the magnetic properties significantly. By hot pressing at 1050° C under 39 MPa, the density of the sample increased from its initial value of 4.0 g/cm^3 up to 5.02 g/cm^3, the coercivity decreased from 5.45 kOe down to 4.8 kOe, and the maximum energy product increased up to 1.4 MGOe. In addition, washing the sample with HCl was reported to improve the saturation magnetization up to 70.7 emu/g and the coercivity up to 6.4 kOe [23].

2.3. SOL-GEL METHODS

Nanosized BaM powders were sucessfully prepared by sol-gel techniques at relatvely low sintering temperatures [22, 34, 35]. Optimization of the experimental conditions necessary to produce high quality ferrite powders, including Fe:Ba ratio, pH, and sintering temperature, was the main concern of published literature in this domain of materials research. These conditions were found to have significant effects on the particle size and magnetic properties of the synthesized hexaferrite powder.

In an investigation of the effects of these conditions, BaM powders were prepared by the sol–gel technique with different Fe:Ba ratio, and sintered at different temperatures between 850° C and 950° C for 1 – 5 h [36]. Powders with particle size below 100 nm were obtained at Fe:Ba = 9 upon sintering at 850° C for 1 h. Optimum saturation magnetization of 70 emu/g and coercivity of 5905 Oe were obtained with Fe:Ba = 10.5, and enhancement of the saturation magnetization up to the theoretical value of 72 emu/g was achieved at lower Fe:Ba ratios. In another study, BaM powder prepared by the sol–gel method from an intitial powder mixture with Fe:Ba = 11.5 and sintered at temperature of 900 – 950° C revealed saturation magnetization of 70 emu/g and coercivity of 5950 Oe [37].

The effects of the ratio of citric acid/metal nitrate (R), Fe:Ba ratio, pH, and annealing temperature on the magnetic properties of BaM powders synthesized by the sol–gel technique revealed that the optimum conditions were $R = 3$, Fe:Ba = 12, and pH = 9 [38]. The variation of the annealing temperature of the powder was found to have little effect on the particle size (~ 40 nm), but significant effect on the magnetic properties, yielding a powder with saturation magnetization of 60.75 emu/g, remanent magnetization of 35.6 emu/g, and coercivity of 5692 Oe at an annealing temperature of 1000° C. Also, a comparable value of the coercivity (5689 Oe) but significantly lower values of the saturation magnetization and remanent magnetization (40.4 emu/g, 22.3 emu/g, respectively) were achieved at a sintering temperature of 900° C by using $R = 2$, Fe:Ba = 12, and pH = 8 [39]. In another study, however, the use of $R = 2$, Fe:Ba = 11.5, and pH = 7 was reported to give optimal magnetic properties at a sintering temperature of 850° C, with saturation magnetization of 55 emu/g, remanent magnetization of 28 emu/g, and coercivity of 5000 Oe [40]. In a yet another study, BaM ferrite prepared at 800° C with R = 2.5, Fe:Ba = 12, and pH = 7 exhibited a saturation magnetization of 74 emu/g and coercivity of 5163 Oe [41]. Further, in a systematic investigation of the influence of the experimental conditions on the BaM phase formation and its magnetic properties revealed that the optimum Fe:Ba ratio was 9 [42]. The highest saturation magnetization of 5750 Oe and saturation magnetization of 67.7 emu/g were achieved with a glycine/metal nitrate molar ratio of 12/9, and calcination temperature of 900° C. In addition, Fe:Ba ratios as low as 4 with R = 1 and pH = 7 were used to synthesize BaM ferrite with improved magnetic properties [43]. The powders were washed with HCl to remove soluble secondary phases and improve the magnetic properties, where a saturation magnetization of 69.4 emu/g, remanent magnetization of 37.5 emu/g, and a coercivivith of 4936 Oe were achieved.

The above discussion of relevant eperimental results indicated the superiority of chemical methods over conventional solid state reaction method in producing hexaferrites with controllable particle size distribution, and improved magnetic properties. Table 1 summarizes the results of representitave research work, where the magnetic parameters of BaM ferrites prepared by different experimental procedures discussed in the previous chapter are listed.

Table 1: Main magnetic parameters of BaM ferrites prepared by different experimental methods.

Method	M_s (emu/g)	M_r (emu/g)	H_c (Oe)	Reference
Conventional Ceramic Method	62	33	3550	[44]
	61	32	2080	[45]
	49	24	1005	[46]
	56		3620	[47]
	51	28	2450	[48]
	57		1943	[49]
	72		4200	[50]
	71	37	4020	[51]
	60		4000	[52]
	70		3470	[53]
Coprecipitation	54		3670	[54]
	64	31	4700	[55]
	72		5340	[56]
	62	37	3646	[57]
	69	36	5440	[58]
	44	23	2604	[31]
	52		4631	[49]
	65		5540	[59]
	70		5044	[22]
	54		6000	[60]
	71		6400	[23]
	65	36	5791	[30]
	68		5720	[20]
	49		4800	[61]

	59	36	1920	[62]
Sol-gel	61	36	5692	[38]
	55		3487	[22]
	70		5905	[36]
	70		5950	[37]
	67		5650	[42]
	38	31	3217	[63]
	55	28	5000	[40]
Auto combustion	64		750	[34]
	50	31	5017	[64]
	71	44	3029	[65]
	60		4250	[43]
	40	22	5689	[39]
	74		5163	[41]
Spray pyrolysis	63	32	5735	[66]
	57		6000	[67]
	40		2500	[68]
Hydrothermal	59	20	1350	[69]
	64		2300	[70]
	65	31	4400	[71]
Microemulsion	49	24	3015	[72]
	64	31	5483	[73]
Molten salt	59		4820	[47]
	72		4650	[74]
	72		4300	[75]

3. EFFECTS OF DIVALENT METAL SUBSTITUTION FOR BARIUM

The variations of the magnetic properties of BaM hexaferrites by adopting different experimental procedures and different experimental conditions are usually associated with the quality and purity of the product, and the level of controlling the particle size and morphology of the magnetic powder. In addition, the range the enhancement of the magnetic properties by adopting different experimental procedures seems to be limited. Accordingly, research work concerning the improvement of the magnetic properties of M-type hexaferrites involved the pursuit of more efficient and economic solutions. Cationic substitution in the standard M-type hexaferrite was found efficient in modifying its magnetic properties, and may involve modifying its intrinsic properties such as the magnetocrystalline anisotropy [76-78].

The most commonly used substituent for Ba in M-type hexaferrite is Sr, with the purpose of improving its properties for permanent magnet applications via the improvement of the coercivity and saturation magnetization [79-81]. The choice of experimental conditions to control the growth of SrM ferrite particles was found to have a dramatic effect on the magnetic properties of the ferrite powder [82-85]. The coercivity of SrM ferrite prepared by the solid-state reaction and double calcination at 1100° C for 4 h and at 1200° C for 2 h exhibited a reduction from 4.36 kOe for Fe:Sr = 12 down to 3.69 kOe for Fe:Sr = 11, and a slight improvement of the saturation magnetization from 72.2 emu/g to 72.6 emu/g [86]. Optimization of the experimental conditions enabled the enhancement of the coercivity of M-type hexaferrites prepared by sol–gel technique from 5905 Oe for BaM up to 6281 Oe for SrM [36]. Also, SrM hexaferrites with coercivity $\geq$ 6.4 kOe were prepared by different experimental procedures with optimized experimental conditions [87-90].

The high temperature required for the formation of M-type hexaferrite is a draw-back for certain device design purposes. Accordingly, lowering the formation temperature of the M-type phase is essential in such cases. This was achieved by adopting variations of the ferrite powder synthesis route. Sol-gel synthesis of SrM ferrite (Fe:Sr = 11) with ethylene glycol as solvent resulted in a powder (SF800) with high coercivity of 6091 Oe, but relatively low saturation magnetization (41.6 emu/g measured at the maximum applied field of 15 kOe) at a calcination temperature of 800° C [91]. An increase in the calcination temperature of the powder up to 1000° C (SF1000) resulted in a significant enhancement of the saturation magnetization up to 62.7 emu/g, and a dramatic reduction of the coercivity down to 1042 Oe. While SF800 could be potentially important for the permanent magnet industry, SF1000 with low coercivity and high saturation magnetization could be potentially important for other applications such as high density magnetic recording. Optimum coercivity of 6.4 kOe and saturation magnetization of 74 emu/g were obtained with SrM (Fe:Sr = 11.6) ferrite prepared by the sol–gel method and sintered at temperatures of 900° C – 950° C [92]. The highest quoted coercivity of SrM ferrite prepared by a modified coprecipitation method at 900° C was 6.8 kOe, and the specific magnetization of the compound at a magnetizing field of 8000 Oe was 54 emu/g [60]. Also, the addition of small amounts of Bi_2O_3 to SrM hexaferrite prepared by ball milling resulted in lowering the sintering temperature, accompanied with a large enhancement of the magnet density, saturation magnetization, and coercivity [93]. The disks made of powders with no additives and sintered at 900° C exhibited a ~ 75% density, saturation magnetization of 33.3 emu/g, and coercivity of 1936 Oe. The addition of 3 wt. % Bi_2O_3 improved the density to 88%, the saturation magnetization to 61.0 emu/g, and the coercivity to 4363 Oe. Sintering the disk at a higher temperature of 1100° C resulted in a further improvement of the density and saturation magnetization up to ~92% and 64.5 emu/g respectively, but with a decline in coercivity down to 3222 Oe.

The production of PbM ($PbFe_{12}O_{19}$) and CaM ($CaFe_{12}O_{19}$) ferrites is of lesser interest due to the inefficiency of replacing Ba^{2+} by Pb^{2+} or Ca^{2+} in improving the magnetic properties of the hexaferrite [94-98]. In a recent article, it was reported that PbM ferrite prepared by the sol–gel method with an acid:metal molar ratio of 1:1 at pH = 7 resulted in the best ferrite with saturation magnetization of 42.28 emu/g and coercivity of 3872 Oe at a sintering temperature of 800° C [99]. In an earlier publication, however, the addition of CaO was reported to improve the coercivity of SrM ferrite [100].

In pursuit of M-type hexaferrites with improved magnetic properties, the effects of substitutions of Ba or Sr by rare-earth metals were also investigated. The substitution of Ba by Eu was found to increase the coercivity from 1.92 kOe for the un-doped sample up to 6.12 kOe for $Ba_{0.75}Eu_{0.25}Fe_{12}O_{19}$ [62]. Also, partial substitution of Ba by a combination of Sr and a RE element (La, Ce, Pr, Nd, and Sm) were found to improve the maximum energy product of the ferrite, culminating at ~ 14% improvement by using 7.5 mole % of $SrCO_3$ and 10 mole % La_2O_3 [101]. However, the substitution of Ba by La was not found to induce any systematic change in the coercivity of BaM hexaferrite [102]. Additionally, Ce-substituted BaM ($Ba_{1-x}Ce_xFe_{12}O_{19}$) revealed improvement of the saturation magnetization in the range of x between 0.05 and 0.15 [61].

The substitution of Sr by La in SrM was reported to improve the magnetic properties of the ferrite at low La concentrations, whereas the coercivity generally decreased with increasing La concentration ($x > 0.3$), and with increasing sintering temperature between 1180° C and 1300° C [103]. Also, in a recent publication, it was demonstrated that both saturation magnetization and coercivity improved upon increasing x in $Sr_{1-x}La_xFe_{12-x}Co_xO_{19}$ [104]. The coercivity, however, did not demonstrate systematic behavior with increasing x, and exhibited a maximum value of 5088 Oe at $x = 0.1$. The substitution of

La-Ce combination in Zn-substituted SrM ferrite, on the other hand, was reported to induce a systematic reduction in both the saturation magnetization and coercivity of the hexaferrite [105]. Further, the effects on the structural and magnetic properties of rare-earth metal substituted SrM ($Sr_{1-x}RE_xFe_{12}O_{19}$) ferrite prepared by the sol–gel method as a function of RE concentration and calcination temperature was recently investigated by Zhou et al. [106]. The substitution of Sr by RE ions was found to result in the development of other oxide phases (α-Fe_2O_3 and $REFeO_3$), accompanied with a general reduction in saturation magnetization. The highest coercivity was achieved by Dy substitution, where at x = 0.5 the coercivity increased from 3282 Oe for SrM to 6717 Oe for $Sr_{0.5}Dy_{0.5}Fe_{12}O_{19}$. This substitution, however, resulted in a reduction of the saturation magnetization from 64.26 emu/g down to only 28.85 emu/g.

The above discussion indicates that limited successes were made in improving the magnetic properties of M-type hexaferrite via the substitution of Ba^{2+} ion by other ions. The need for further improvements of the magnetic properties to satisfy the requirements of different technological and industrial applications instigated vigorous research work involving the substitution of Fe^{3+} ions by trivalent metal ions, or by combinations of divalent and tetravalent metal ions. The effects of such substitutions will be the subject of the following subsections.

4.   TRIVALENTAL METAL SUBSTITUTION FOR IRON

A great deal of research work had been carried out with the purpose of improving the magnetic properties of BaM by substituting Fe^{3+} by a trivalent magnetic or nonmagnetic metal ions such as Al, Ga, Sc, In, Cr, Mn, and Co. The magnetic properties of the substituted ferrites are influenced by both the ordering of the substituent ions in the various interstitial sites of the hexaferrite lattice, and the level of the substitution. Generally speaking, the substitution of iron ions by nonmagnetic ions at spin-up sites results in a decrease in the saturation magnetization. The substitution at spin-down sites, however, is expected to improve the saturation magnetization due to the enhancement of the net magnetic moment per unit cell. Such substitution, however, results in the decrease of the spin-down moment, and the consequent weakening of the superexchange interaction responsible for the collinear magnetic structure in the uniaxial ferrite. The effect of this reduction of the strength of the superexchange interaction could entail deviations from spin collinearity [76], with the consequent reduction in the saturation magnetization due to spin canting. Also, this reduction in superexchange strength is usually accompanied by the decrease in the Curie temperature of the ferrite. The competition between effects leading to the enhancement of the net moment per cell, and those resulting in the reduction of the strength of the superexchange interactions, results in the final magnetic state of the material.

Aluminum substitution for iron in M-type hexaferrite is probably the most effective in enhancing the coercivity, accompanied by an unfavorable reduction of the saturation magnetization [45, 50, 65, 107, 108]. Since the coercivity is influenced by the particle size, a reduction in coercivity is normally observed at high sintering temperatures due to particle growth beyond the critical single domain size. Al-substituted BaM ferrites ($BaFe_{10}Al_2O_{19}$) were prepared by ball milling starting powders with (Fe + Al):Ba ratio of 11 and sintering at different temperatures to investigate the effect of heat treatment on the magnetic properties of the ferrite [108]. The samples sintered at temperatures 1000–1100° C revealed the presence of a small fraction of α-Fe_2O_3 oxide phase. At a sintering temperature of 1200° C, the saturation magnetization of the ferrite decreased from ~ 60 emu/g for the unsubstituted sample down to ~ 34 emu/g for the Al-substituted sample. The highest coercivity of 9.3 kOe was obtained upon sintering $BaFe_{10}Al_2O_{19}$ at 1100° C, with a saturation magnetization of 36 emu/g. On the other hand, the effect of Al-substitution on the magnetic properties of SrM ferrites ($SrFe_{12-x}Al_xO_{19}$) prepared by sol–gel auto-combustion, using stoichiometric metal nitrate ratios and citric acid/nitrate ratio of 1 at pH = 6.5 was investigated [109]. The saturation magnetization of powders sintered at 950° C decreased systematically from 60.0 emu/g for $x = 0$ down to 11.8 emu/g at $x = 4$. The coercivity of $SrFe_8Al_4O_{19}$, however, demonstrated a record coercivity of 16.2 kOe. However, contradictory results on Al-doped BaM were reported elsewhere [110]. Further, the coercivity of Al-substituted SrM ferrites was improved by partial substitution of Sr by a rare-earth (RE) element [111]. A systematic study of the effect of the type of RE ion substitution on the magnetic properties of $Sr_{0.9}RE_{0.1}Fe_{10}Al_2O_{19}$ revealed a significant improvement of the coercivity of SrM ferrite, Pr being the most effective in enhancing the coercivity from 7.4 kO up to 11.0 kOe [112]. At this point, it is worth mentioning that the Al substitution results in unwanted reduction of the saturation magnetization. Such substitution can therefore be useful in applications where cheap, high coercivity materials are sought. Also, such materials are important for microwave device applications at ultrahigh frequency, as will be explained in the next chapter. Particularly, Al-substituted SrM hexaferrites with $0 \leq x \leq 1.9$ exhibited an increase in magnetocrystalline anisotropy field from 19 kOe to 34 kOe, thus extending the applicability of these ferrites in constructing microwave devices operating at ultrahigh frequencies in he range 60 – 90 GHz [113].

The effect of Cr^{3+} substitution for Fe^{3+} on the magnetic properties of M-type hexaferrite was investigated. Cr-substituted BaM hexaferrites ($BaFe_{12-x}Cr_xO_{19}$) were prepared by the sol sol-gel auto-combustion method with citric acid/metal nitrate ratio of 2 at pH = 8; the powder was subsequently sintered at 900° C for 8 h [39]. Only the pure sample was composed of a pure hexaferrite phase, whereas all substituted samples contained significant amounts of α-Fe$_2$O$_3$ oxide phase. The magnetization measured at the maximum applied field of 9 kOe decreased systematically from 40.4 emu/g to 5.1 emu/g with increasing x from 0.00 to 1.00, and the coercivity decreased in a slower rate from 5689 Oe to 5396 Oe. In light of the high coercivity of the substituted ferrites, one can conclude that the magnetic phase in the samples is BaM, and the sharp decrease in saturation magnetization is due to the development of the α-Fe$_2$O$_3$ oxide phase, which becomes dominant at $x \geq 0.75$. In another study, Cr-substituted BaM ferrites were prepared under similar conditions, except the powders were calcined at 1100° C, ground, and then sintered at 1200° C [114]. In this study, BaM pure phase was obtained at Cr-concentrations up to $x = 0.8$, and the coercivity increased from $\sim$ 2.0 kOe at $x = 0.0$ to $\sim$ 5.2 kOe at $x = 0.8$. The decrease of the saturation magnetization (evaluated from the law of approach to saturation) from $\sim$69 emu/g at $x = 0.0$ to $\sim$ 43 emu/g was associated with magnetic dilution or the development of non-collinear magnetic structure instigated by the substitution of Cr^{3+} ions for Fe^{3+} ions at 2a and 12k spin-up sublattices. Considering the results of these two studies, one may conclude that the presence of the α-Fe$_2$O$_3$ oxide phase in Cr-substituted BaM is due to an incomplete solid state reaction at low temperature, and that sintering temperatures > 1100° C are required even in the sol-gel route which normally requires lower hexaferrite formation temperatures.

Chromium substitution for iron in SrM ferrites (SrFe$_{12-x}$Cr$_x$O$_{19}$) prepared by conventional solid state reaction of ball-milled powders with stoichiometric Fe:Sr ratios was also found to develop α-Fe$_2$O$_3$ oxide at $x > 0.3$, even at sintering temperatures as high as 1220° C [115]. Optimization of the experimental conditions resulted in only little improvement of the magnetic properties of the compound with $x = 0.3$, where the coercivity remained $\gtrsim$ 4000 Oe. In addition, Cr-substituted SrM ferrite prepared by hydrothermal synthesis and microwave sintering at 950° C revealed the presence of only small fractions of α-Fe$_2$O$_3$ at $x \geq 0.5$ [116, 117]. Although the saturation magnetization of the compound with $x = 0.9$ was found rather low (30 emu/g), the coercivity was as high as 7335 Oe.

In a recent study, it was shown that vanadium substitution for iron in BaM ferrites prepared by ball milling and sintering at different temperatures leads to significant effects on the magnetic properties [118]. The coercivity of BaFe$_{12-x}$V$_x$O19 sintered at 1100° C was found to increase from 3500 Oe for the unsubstituted sample up to 4400 Oe at $x = 0.2$, with a decrease of the saturation magnetization from 69.0 emu/g down to 56.8 emu/g, and of the remanent magnetization from 37.3 emu/g to 30.0 emu/g. The increase in sintering temperature up to 1300° C resulted in an increase in saturation magnetization from 21.8 emu/g to 26.6 emu/g, and a decrease in coercivity from 4.0 kOe down 1.6 kOe for the sample with $x = 0.5$. The addition of extra Ba in the starting powder (for (Fe+V):Ba = 6.3) resulted in an increase in the fraction of BaM at this higher end of the vanadium substitution, and a consequent significant increase of the saturation magnetization from 21.8 emu/g up to 56.3 emu/g, and remanent magnetization from 11.4 emu/g to 29.3 emu/g at a sintering temperature of 1200° C. The coercivity of the sample, however, dropped to 2.1 kOe due to the elevated temperature treatment. Further improvement of the saturation magnetization up to 59.6 emu/g, and remanent magnetization up to 32.8 emu/g, without any noticeable effect on the coercivity was achieved by washing the powder with HCl solution to remove the soluble nonmagnetic phases.

The substitution of Fe by Mn in BaM prepared by ball milling and sintering at 1050° C was found to exhibit a linear increase of the coercivity from about 4600 Oe to about 5100 Oe with increasing Mn concentration from $x = 0.0$ to $x = 0.5$ [119]. The saturation magnetization, however, decreased from about 63 emu/g to about 57 emu/g in this concentration range. It was argued that the magnetocrystalline anisotropy does not change with Mn substitution, and the increase in coercivity was associated with the reduction of the crystallite size. The crystallite size in the range $x = 0.2 - 0.5$, however, did not change systematically and appreciably, which would raise doubts around the interpretation of the coercivity behavior in terms of crystallite size. Alternative reasons behind this behavior were suggested, including pinning effect and Fe–Mn–Fe superexchange interactions. An opposite effect of Mn substitution on the coercivity of ferrites prepared by the conventional ceramic method [120], and by the sol–gel auto-combustion method [121] were reported.

The substitution of Fe^{3+} by Gd^{3+} was also reported to improve the saturation magnetization and the coercivity of BaM ferrite. $BaFe_{12-x}Gd_xO_{19}$ nanopowders prepared by the sol–gel technique and sintered at 1100° C for 4 h exhibited a coercivity of 6202 Oe at $x = 0.1$, and the saturation magnetization increased up to 81.34 emu/g at $x = 0.3$ [122]. In this study, a small diffraction peak between the (107) and (114) main reflections of the BaM structure was marked and argued to be an indication of the development of $\alpha\text{-}Fe_2O_3$ phase at $x = 0.2$ and 0.3. The presence of such nonmagnetic secondary phase, however, should have the effect of reducing the saturation magnetization, contrary to observation. In addition, the absence of the main structural peak of $\alpha\text{-}Fe_2O_3$ at an angular position around 35.6° is an indication that the marked peak does not correspond to the nonmagnetic iron oxide phase, but rather, possibly to another spinel phase. On the other hand, the substitution of Fe^{3+} ions by Ce^{3+} ions in BaM hexaferrites prepared by the sol–gel method and annealed at 1000° C for 5 h was reported to result in a decrease of the saturation magnetization, and an increase in coercivity of the ferrite [123]. The figures quoted for the coercivity in this study (18.99 kOe to 23.05 kOe) are, however, unrealistic.

Table 2: Some of the highest magnetic properties reported on M-type hexaferrites prepared by different techniques under optimized experimental conditions. The sintering temperature T_S is shown.

Compound	Method	T_S (°C)	M_s (emu/g)	H_c(Oe)	Reference
$BaFe_{12}O_{19}$	Conventional Ceramic	1000	52.3	4537	[19]
$BaFe_{12}O_{19}$	Coprecipitation	1100	70.7	6400	[23]
$SrFe_{12}O_{19}$	Sol–Gel	900-950	74	6400	[92]
$SrFe_{12}O_{19}$	Sol–Gel	800	41.6	6091	[91]
$BaFe_{12}O_{19}$	Coprecipitation– Molten salt	800-850	54	6000	[60]
$SrFe_{12}O_{19}$	Coprecipitation– Molten salt	900	54	6800	[60]
$SrFe_{11.1}Cr_{0.9}O_{19}$	Hydrothermal	950	30	7335	[116, 117]
$BaFe_{10}Al_2O_{19}$	Conventional Ceramic	1100	36	9300	[108]
$SrFe_{10}Al_2O_{19}$	Sol–Gel	1100	36.5	7400	[112]
$SrFe_8Al_4O_{19}$	Sol-Gel	950	11.8	16200	[109]
$Sr_{0.9}Y_{0.1}Fe_{10}Al_2O_{19}$	Sol–Gel	1100	36.0	8870	[112]
$Sr_{0.9}La_{0.1}Fe_{10}Al_2O_{19}$	Sol–Gel	1100	33.7	88100	[112]
$Sr_{0.9}Ce_{0.1}Fe_{10}Al_2O_{19}$	Sol–Gel	1100	32.7	9480	[112]
$Sr_{0.9}Pr_{0.1}Fe_{10}Al_2O_{19}$	Sol–Gel	1100	30.8	11000	[112]
$Sr_{0.9}Nd_{0.1}Fe_{10}Al_2O_{19}$	Sol–Gel	1100	34.0	10000	[112]
$Sr_{0.9}Sm_{0.1}Fe_{10}Al_2O_{19}$	Sol–Gel	1100	33.6	9340	[112]
$Sr_{0.9}Gd_{0.1}Fe_{10}Al_2O_{19}$	Sol–Gel	1100	33.9	9620	[112]

The experimental results reviewed above indicated effectiveness of different metal substitution strategies, and modifications of the experimental conditions in improving the magnetic properties, especially the coercivity, of M-type hexaferrites. Some of the reported magnetic data of M-type ferrites with particularly high coercivities are listed in Table 2.

5.   DIVALENT-TETRAVALENT METAL SUBSTITUTIONS FOR IRON

Generally, the substitution of Fe^{3+} ions by combinations of divalent-tetravalent metal ions result in a reduction of the coercivity of the hexaferrite, which is a negative effect for hard permanent magnet applications. For practical purposes, however, a coercivity $\leq$ 2000 Oe is one of the main requirements of high density magnetic recording applications. The main challenge in fabricating magnetic materials with suitable properties, however, resides in tuning the coercivity, controlling the particle size distribution, and maintaining the saturation magnetization as high as possible, which instigated intensive research work aimed at modifying the magnetic characteristics and particle morphology of M-type hexaferrites for parallel and perpendicular magnetic recording applications [124-134]. The substitution of Fe^{3+} ions by special combinations of ions was found efficient in providing the necessary modifications for magnetic recording applications. In addition, the substitution could modify the magnetocrystalline anisotropy field, thus providing an efficient tool for tuning the magnetic anisotropy in order to shift the natural resonance frequency to the value required for the design of an electromagnetic microwave absorber [135-137]. Accordingly, extensive research work was devoted to the synthesis and characterization of M-type hexaferrites with Fe^{3+} ions substituted by different metal ion combinations [24, 138-154].

In particular, Co–Ti substituted hexaferrites ($BaFe_{12-2x}Co_xTi_xO_{19}$) received considerable interest due to the effectiveness of this combination in reducing the coercivity down to values suitable for magnetic recording and data storage media, and microwave devices operating in the hyper-frequency range [134, 136, 152, 155-170]. Specifically, Co–Ti substituted BaM ferrite prepared by the sol–gel method exhibited a noticeable reduction in coercivity down to ~ 1400 – 2400 Oe at $0.4 \leq x \leq 0.6$, with only few percent decrease in saturation magnetization [134]. Also, Co–Ti substituted BaM ferrites prepared by a modified ceramic method exhibited a coercivity in the range ~ 1180 – 2166 Oe, with a slight improvement of the saturation magnetization for x values in the range 0.60 – 0.75. In addition, relatively high saturation magnetization and coercivity in the range of ~1000 – 2000 Oe was obtained for Co–Ti substitution in the range $x = 0.50 – 0.75$ [124]. It should be noted, however, that the mean particle size plays an important role in the coercivity of the ferrite. Powders with mean particle size exceeding the critical single domain size are normally composed of multi-domain particles, where the magnetization by domain wall motion results in an unfavorable reduction of the coercivity. A demonstration of the effect of particle size on the coercivity was provided by the work of Muller et al. on $BaFe_{10.48}Co_{0.76}Ti_{0.76}$ prepared by glass crystallization method [15]. Fig. 2 is a reproduction of the data, which clearly shows the monotonic growth of the mean particle size with annealing temperature. The coercivity of single domain particles ($D < 0.5$ μm) is not affected appreciably, and remains around the typical value of ~ 1200 Oe for the composition adopted in the study. The sharp drop in coercivity of compounds with $D > 0.5$ μm is apparently due to the development of domain walls within the individual particles. Further, Gruskova et al. reported a coercivity of ~ 1070 Oe with improved saturation magnetization for $x = 0.5$ in Co–Ti substituted BaM ferrite [157]. For this composition, however, available literature indicates that the coercivity of a single-domain powder should be $\geq$ 2000 Oe, almost double the value reported by Gruskova et al. The observed unusually low coercivity could therefore be associated with the presence of multi-domain particles in the sample, based on electron microscopy imaging which reveled particle size > 0.5 μm.

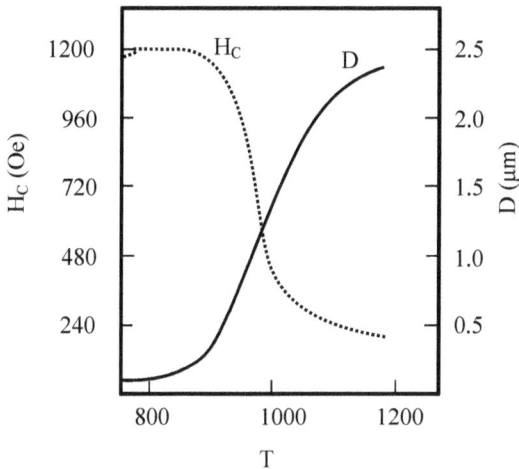

Fig. 2. Dependence of the mean particle size (D) and intrinsic coercivity (H_C) of $BaFe_{10.48}Co_{0.76}Ti_{0.76}$ ferrites on the annealing temperature (reproduced from the data in Ref. [15])

Other metal combination substitutions were also found suitable for the production of materials with the desired characteristics for magnetic recording applications [133].

Specifically, Co–Zn substitution was reported to improve the characteristics of BaM for magnetic recording applications by improving the saturation magnetization and reducing the switching field distribution, which is an important parameter for the performance of the magnetic recording material [171]. Also, Zr–Cd substituted SrM ferrite exhibited properties suitable for high density magnetic recording, with marginal decrease of the saturation magnetization up to $x = 0.4$ [172]. However, Cu-V substitution revealed an improvement of the saturation magnetization from 69.0 emu/g for the un-substituted sample up to 72.6 emu/g at $x = 0.1$, with a decrease in coercivity down to ~ 1635 Oe [173]. In addition, $BaFe_{1-x}(Mg_{0.5}Ti_{0.5})_xO_{19}$ revealed reduction in saturation to values required for magnetic recording, with only little effect on the saturation magnetization up to $x = 2$ [77].

Although most substitutions in the above literature were generally accompanied by some degree of decline in saturation magnetization, some substitution strategies, such as Ti–Ru substitution in BaM [52] and SrM [143] ferrites, and Zn-Ru and Ni-Ru substitution in BaM ferrite [142] exhibited a significant improvement of the saturation magnetization, accompanied with the reduction of the coercivity down to levels demanded by magnetic recording requirements. Also, $BaFe_{12-2x}Mn_xTi_xO_{19}$ exhibited an improvement in saturation accompanied with a reduction in coercivity down to 1850 Oe at $x = 1.0$, and a decrease in both the saturation magnetization and coercivity at higher concentrations [174]. In addition, Co-Zr substitution for Fe^{3+} ions in BaM ferrite prepared by citrate precursor method resulted in an improvement in saturation magnetization, and a reduction in coercivity down to ~ 1630 Oe at $x = 0.6$ and sintering temperature of 1070° C [175]. In another study, the substitution of Fe^{3+} by Co–Zr in BaM ferrite prepared by citrate precursor method also exhibited a reduction in coercivity down to 1985 Oe with saturation magnetization of 55 emu/g at $x = 0.6$ [176]. Further, the substitution of Fe by Co–Sn in coprecipitated BaM powder resulted in a significant increase in saturation magnetization up to 82.0 emu/g at $x = 0.5$, and 84.3 emu/g at $x = 0.75$, accompanied by a reduction in coercivity down to 2002 Oe and 1012 Oe, respectively[24].

Co–Sn substituted BaM prepared by reverse microemulsion technique also revealed reduction of the coercivity down to 1510 Oe at $x = 0.5$ and 854 Oe at $x = 0.7$, but the saturation magnetization decreased from 70.4 emu/g at $x = 0.5$ down to 62.0 emu/g at $x = 0.7$ [177]. The reduction in saturation magnetization at high substitution levels in the latter study was correlated with the lower magnetic moments of Co^{2+} and Sn^{4+} ions in comparison with Fe^{3+} ions, and their preferred site of occupation.

In the course of reviewing the relevant experimental results concerning the magnetic properties of M-type hexaferrites, we have noticed some general trends of the behaviors of the coercivity and saturation magnetization of the ferrites. Contradictory results, however, were reported by different researchers adopting different experimental strategies. These contradictory results may reflect the sensitivity of the properties of the final product to the method of powder synthesis, and adopted experimental procedures and conditions. The reproducibility in synthesizing M-type hexaferrites with specific magnetic properties, therefore, relies on adopting fixed experimental strategies and conditions.

REFERENCES

[1] R.C. Pullar, Hexagonal ferrites: a review of the synthesis, properties and applications of hexaferrite ceramics, Progress in Materials Science, 57 (2012) 1191-1334.
http://dx.doi.org/10.1016/j.pmatsci.2012.04.001

[2] V.G. Harris, A. Geiler, Y. Chen, S.D. Yoon, M. Wu, A. Yang, Z. Chen, P. He, P.V. Parimi, X. Zuo, Recent advances in processing and applications of microwave ferrites, Journal of Magnetism and Magnetic Materials, 321 (2009) 2035-2047.
http://dx.doi.org/10.1016/j.jmmm.2009.01.004

[3] Ü. Özgür, Y. Alivov, H. Morkoç, Microwave ferrites, part 1: fundamental properties, Journal of Materials Science: Materials in Electronics, 20 (2009) 789-834.
http://dx.doi.org/10.1007/s10854-009-9923-2

[4] Ü. Özgür, Y. Alivov, H. Morkoç, Microwave ferrites, part 2: passive components and electrical tuning, Journal of Materials Science: Materials in Electronics, 20 (2009) 911-952.
http://dx.doi.org/10.1007/s10854-009-9924-1

[5] G. Turilli, F. Licci, S. Rinaldi, A. Deriu, Mn^{2+}, Ti^{4+} substituted barium ferrite, Journal of Magnetism and Magnetic Materials, 59 (1986) 127-131.
http://dx.doi.org/10.1016/0304-8853(86)90019-3

[6] J. Smit, H.P.J. Wijn, Ferrites, Wiley, New York, 1959.

[7] I.Y. Gershov, Properties and uses of barium ferrite ceramic magnets, Powder Metallurgy and Metal Ceramics, 2 (1963) 227-234.
http://dx.doi.org/10.1007/BF00774277

[8] I.Y. Gershov, Barium ferrite permanent magnets, Soviet Powder Metallurgy and Metal Ceramics, 1 (1964) 386-393.
http://dx.doi.org/10.1007/BF00774124

[9]   L.Y. Mondin, Effect of heat treatment on the magnetic properties of barium ferrite powders, Powder Metallurgy and Metal Ceramics, 8 (1969) 419-422.
http://dx.doi.org/10.1007/BF00776620

[10]   D. Chen, Y. Liu, Y. Li, W. Zhong, H. Zhang, Low-temperature sintering of M-type barium ferrite with $BaCu(B_2O_5)$ additive, Journal of Magnetism and Magnetic Materials, 324 (2012) 449-452.
http://dx.doi.org/10.1016/j.jmmm.2011.08.016

[11]   U. Topal, Improvement of the remanence properties and the weakening of interparticle interactions in $BaFe_{12}O_{19}$ particles by B_2O_3 addition, Physica B: Condensed Matter, 407 (2012) 2058-2062.
http://dx.doi.org/10.1016/j.physb.2012.02.004

[12]   U. Topal, Towards Further Improvements of the Magnetization Parameters of B_2O_3-Doped $BaFe_{12}O_{19}$ Particles: Etching with Hydrochloric Acid, Journal of superconductivity and novel magnetism, 25 (2012) 1485-1488.
http://dx.doi.org/10.1007/s10948-012-1486-4

[13]   N. Vitkina, Y.M. Vernigorov, B. Gasanov, Influence of heat treatment on the structure and properties of oxide magnets produced by "dry" pressing, Soviet Powder Metallurgy and Metal Ceramics, 29 (1990) 23-25.
http://dx.doi.org/10.1007/BF00796087

[14]   D. Martínez, J. Rivas, F. Walz, M. Gayoso, C. Rodríguez, M. Señarís, Influence of non-stoichiometry on magnetic relaxation in polycrystalline barium ferrites, Journal of Magnetism and Magnetic Materials, 83 (1990) 468-470.
http://dx.doi.org/10.1016/0304-8853(90)90589-1

[15]   R. Müller, H. Pfeiffer, W. Schüppel, Variation of the magnetic properties of barium ferrite powders by heat treatment, Journal of magnetism and magnetic materials, 101 (1991) 237-238.
http://dx.doi.org/10.1016/0304-8853(91)90742-S

[16]   Y. Goto, T. Takada, Phase Diagram of the System $BaO\text{-}Fe_2O_3$, Journal of the American Ceramic Society, 43 (1960) 150-153.
http://dx.doi.org/10.1111/j.1151-2916.1960.tb14330.x

[17]   Francisco Carmona, Aurelio Martin Blanco, Carlos Alemany, Magnetic viscosity in Ba-ferrite, Journal of Magnetism and Magnetic Materials, 92 (1991) 417-423.
http://dx.doi.org/10.1016/0304-8853(91)90856-6

[18]   S. Campbell, W. Kaczmarek, G.-M. Wang, Ball-milled barium ferrite-effects of annealing, Nanostructured Materials, 6 (1995) 687-690.
http://dx.doi.org/10.1016/0965-9773(95)00151-4

[19] D. Lisjak, The low-temperature sintering of M-type hexaferrites, Journal of the European Ceramic Society, 32 (2012) 3351-3360.
http://dx.doi.org/10.1016/j.jeurceramsoc.2012.04.003

[20] P. Görnert, W. Schüppel, E. Sinn, F. Schumacher, K.A. Hempel, G. Turilli, A. Paoluzi, M. Rösler, Comparative measurements of the effective anisotropy field H_a for barium ferrites, Journal of magnetism and magnetic materials, 114 (1992) 193-201.
http://dx.doi.org/10.1016/0304-8853(92)90344-N

[21] S.R. Janasi, D. Rodrigues, F.J. Landgraf, M. Emura, Magnetic properties of coprecipitated barium ferrite powders as a function of synthesis conditions, Magnetics, IEEE Transactions on, 36 (2000) 3327-3329.
http://dx.doi.org/10.1109/intmag.2000.872434

[22] J.-P. Wang, L. Ying, M.-L. Zhang, Y.-j. QIAO, X. Tian, Comparison of the Sol-gel Method with the Coprecipitation Technique for Preparation of Hexagonal Barium Ferrite, Chemical Research in Chinese Universities, 24 (2008) 525-528.
http://dx.doi.org/10.1016/S1005-9040(08)60110-5

[23] W. Roos, H. Haak, C. Voigt, K. Hempel, Microwave absorption and static magnetic properties of coprecipitated barium ferrite, Le Journal de Physique Colloques, 38 (1977) C1-35-C31-37.

[24] S. Nilpairach, W. Udomkichdaecha, I. Tang, Coercivity of the co-precipitated prepared hexaferrites, $BaFe_{12-2x}Co_xSn_xO_{19}$, Journal of the Korean Physical Society, 48 (2006) 939-945.

[25] P. Garcia-Casillas, A. Beesley, D. Bueno, J. Matutes-Aquino, C. Martinez, Remanence properties of barium hexaferrite, Journal of alloys and compounds, 369 (2004) 185-189.
http://dx.doi.org/10.1016/j.jallcom.2003.09.100

[26] A. Martin, F. Carmona, Intrinsic coercive force variation with packing in fine particle assemblies, Magnetics, IEEE Transactions on, 4 (1968) 259-262.

[27] D.-H. Chen, Y.-Y. Chen, Synthesis of barium ferrite ultrafine particles by coprecipitation in the presence of polyacrylic acid, Journal of colloid and interface science, 235 (2001) 9-14.
http://dx.doi.org/10.1006/jcis.2000.7340

[28] D. Lisjak, M. Drofenik, The mechanism of the low-temperature formation of barium hexaferrite, Journal of the European Ceramic Society, 27 (2007) 4515-4520.

http://dx.doi.org/10.1016/j.jeurceramsoc.2007.02.202

[29] S.E. Jacobo, L. Civale, M.A. Blesa, Evolution of the magnetic properties during the thermal treatment of barium hexaferrite precursors obtained by coprecipitation from barium ferrate (VI) solutions, Journal of magnetism and magnetic materials, 260 (2003) 37-41.
http://dx.doi.org/10.1016/S0304-8853(01)00924-6

[30] J. Matutes-Aquino, S. Diaz-Castanón, M. Mirabal-Garcia, S. Palomares-Sánchez, Synthesis by coprecipitation and study of barium hexaferrite powders, Scripta materialia, 42 (2000) 295-299.
http://dx.doi.org/10.1016/S1359-6462(99)00350-4

[31] M. Rashad, I. Ibrahim, A novel approach for synthesis of M-type hexaferrites nanopowders via the co-precipitation method, Journal of Materials Science: Materials in Electronics, 22 (2011) 1796-1803.
http://dx.doi.org/10.1007/s10854-011-0365-2

[32] P. Shepherd, K.K. Mallick, R.J. Green, Magnetic and structural properties of M-type barium hexaferrite prepared by co-precipitation, Journal of magnetism and magnetic materials, 311 (2007) 683-692.
http://dx.doi.org/10.1016/j.jmmm.2006.08.046

[33] H.B. von Basel, K.A. Hempel, Static magnetic properties of pressure-sintered barium ferrite, Journal of Magnetism and Magnetic Materials, 38 (1983) 316-318.
http://dx.doi.org/10.1016/0304-8853(83)90373-6

[34] V.C. Chavan, S.E. Shirsath, M.L. Mane, R.H. Kadam, S.S. More, Transformation of hexagonal to mixed spinel crystal structure and magnetic properties of Co^{2+} substituted $BaFe_{12}O_{19}$, Journal of Magnetism and Magnetic Materials, 398 (2016) 32-37.
http://dx.doi.org/10.1016/j.jmmm.2015.09.002

[35] D. Suastiyanti, A. Sudarmaji, B. Soegijono, Influence of Ba/Fe mole ratios on magnetic properties, crystallite size and shifting of X-ray diffraction peaks of nanocrystalline $BaFe_{12}O_{19}$ powder, prepared by sol gel auto combustion, International Conference on Physics and its Applications, AIP Publishing, 2012, pp. 238-241.

[36] C. Sürig, D. Bonnenberg, K. Hempel, P. Karduck, H. Klaar, C. Sauer, Effects of Variations in Stoichiometry on M-Type Hexaferrites, Le Journal de Physique IV, 7 (1997) C1-315-C311-316.

[37] W. Zhong, W. Ding, N. Zhang, J. Hong, Q. Yan, Y. Du, Key step in synthesis of ultrafine $BaFe_{12}O_{19}$ by sol-gel technique, Journal of Magnetism and Magnetic Materials, 168 (1997) 196-202.
http://dx.doi.org/10.1016/S0304-8853(96)00664-6

[38] Y. Li, Q. Wang, H. Yang, Synthesis, characterization and magnetic properties on nanocrystalline $BaFe_{12}O_{19}$ ferrite, Current applied physics, 9 (2009) 1375-1380.
http://dx.doi.org/10.1016/j.cap.2009.03.002

[39] V.N. Dhage, M. Mane, M. Babrekar, C. Kale, K. Jadhav, Influence of chromium substitution on structural and magnetic properties of $BaFe_{12}O_{19}$ powder prepared by sol–gel auto combustion method, Journal of Alloys and Compounds, 509 (2011) 4394-4398.
http://dx.doi.org/10.1016/j.jallcom.2011.01.040

[40] D. Bahadur, S. Rajakumar, A. Kumar, Influence of fuel ratios on auto combustion synthesis of barium ferrite nano particles, Journal of chemical sciences, 118 (2006) 15-21.
http://dx.doi.org/10.1007/BF02708761

[41] M. Han, Y. Ou, W. Chen, L. Deng, Magnetic properties of Ba-M-type hexagonal ferrites prepared by the sol–gel method with and without polyethylene glycol added, Journal of alloys and compounds, 474 (2009) 185-189.
http://dx.doi.org/10.1016/j.jallcom.2008.06.047

[42] Y. Meng, M. He, Q. Zeng, D. Jiao, S. Shukla, R. Ramanujan, Z. Liu, Synthesis of barium ferrite ultrafine powders by a sol–gel combustion method using glycine gels, Journal of Alloys and Compounds, 583 (2014) 220-225.
http://dx.doi.org/10.1016/j.jallcom.2013.08.156

[43] H. Sözeri, Z. Durmuş, A. Baykal, E. Uysal, Preparation of high quality, single domain $BaFe_{12}O_{19}$ particles by the citrate sol–gel combustion route with an initial Fe/Ba molar ratio of 4, Materials Science and Engineering: B, 177 (2012) 949-955.
http://dx.doi.org/10.1016/j.mseb.2012.04.023

[44] M. Manawan, A. Manaf, B. Soegijono, A. Yudi, Microstructural and magnetic properties of Ti^{2+}-Mn^{4+} substituted barium hexaferrite, Advanced Materials Research, 896 (2014) 401-405.
http://dx.doi.org/10.4028/www.scientific.net/AMR.896.401

[45] S. El-Sayed, T. Meaz, M. Amer, H. El Shersaby, Magnetic behavior and dielectric properties of aluminum substituted M-type barium hexaferrite, Physica B: Condensed Matter, 426 (2013) 137-143.

http://dx.doi.org/10.1016/j.physb.2013.06.026

[46] V.V. Soman, V. Nanoti, D. Kulkarni, Dielectric and magnetic properties of Mg–Ti substituted barium hexaferrite, Ceramics International, 39 (2013) 5713-5723. *http://dx.doi.org/10.1016/j.ceramint.2012.12.089*

[47] S. Dursun, R. Topkaya, N. Akdoğan, S. Alkoy, Comparison of the structural and magnetic properties of submicron barium hexaferrite powders prepared by molten salt and solid state calcination routes, Ceramics International, 38 (2012) 3801-3806. *http://dx.doi.org/10.1016/j.ceramint.2012.01.028*

[48] U. Topal, A simple synthesis route for high quality $BaFe_{12}O_{19}$ magnets, Materials Science and Engineering: B, 176 (2011) 1531-1536. *http://dx.doi.org/10.1016/j.mseb.2011.09.019*

[49] H. Sözeri, Effect of pelletization on magnetic properties of $BaFe_{12}O_{19}$, Journal of Alloys and Compounds, 486 (2009) 809-814. *http://dx.doi.org/10.1016/j.jallcom.2009.07.072*

[50] I. Bsoul, S. Mahmood, Structural and magnetic properties of $BaFe_{12-x}Al_xO_{19}$ prepared by milling and calcination, Jordan Journal of Physics, 2 (2009) 171-179.

[51] I. Bsoul, S. Mahmood, Magnetic and structural properties of $BaFe_{12-x}Ga_xO_{19}$ nanoparticles, Journal of Alloys and Compounds, 489 (2010) 110-114. *http://dx.doi.org/10.1016/j.jallcom.2009.09.024*

[52] I. Bsoul, S. Mahmood, A.-F. Lehlooh, Structural and magnetic properties of $BaFe_{12-2x}Ti_xRu_xO_{19}$, Journal of Alloys and Compounds, 498 (2010) 157-161. *http://dx.doi.org/10.1016/j.jallcom.2010.03.142*

[53] S. Mahmood, A. Aloqaily, Y. Maswadeh, A. Awadallah, I. Bsoul, H. Juwhari, Structural and Magnetic Properties of Mo-Zn Substituted $(BaFe_{12-4x}Mo_xZn_{3x}O_{19})$ M-Type Hexaferrites, Material Science Research India, 11 (2014) 09-20.

[54] R.S. Alam, M. Moradi, H. Nikmanesh, J. Ventura, M. Rostami, Magnetic and microwave absorption properties of $BaMg_{x/2}Mn_{x/2}Co_xTi_{2x}Fe_{12-4x}O_{19}$ hexaferrite nanoparticles, Journal of Magnetism and Magnetic Materials, 402 (2016) 20-27. *http://dx.doi.org/10.1016/j.jmmm.2015.11.038*

[55] H.-F. Yu, $BaFe_{12}O_{19}$ powder with high magnetization prepared by acetone-aided coprecipitation, Journal of Magnetism and Magnetic Materials, 341 (2013) 79-85. *http://dx.doi.org/10.1016/j.jmmm.2013.04.030*

[56]   Y. Liu, M.G. Drew, Y. Liu, Preparation and magnetic properties of barium ferrites substituted with manganese, cobalt, and tin, Journal of Magnetism and Magnetic Materials, 323 (2011) 945-953.
http://dx.doi.org/10.1016/j.jmmm.2010.11.075

[57]   M. Rashad, I. Ibrahim, Improvement of the magnetic properties of barium hexaferrite nanopowders using modified co-precipitation method, Journal of Magnetism and Magnetic Materials, 323 (2011) 2158-2164.
http://dx.doi.org/10.1016/j.jmmm.2011.03.023

[58]   E. Pashkova, E. Solovyova, I. Kotenko, T. Kolodiazhnyi, A. Belous, Effect of preparation conditions on fractal structure and phase transformations in the synthesis of nanoscale M-type barium hexaferrite, Journal of Magnetism and Magnetic Materials, 323 (2011) 2497-2503.
http://dx.doi.org/10.1016/j.jmmm.2011.05.026

[59]   G. Litsardakis, I. Manolakis, C. Serletis, K. Efthimiadis, High coercivity Gd-substituted Ba hexaferrites, prepared by chemical coprecipitation, Journal of Applied Physics, 103 (2008) 07E501.

[60]   Z. Zong-yu, G. Bi-jun, M. Xue-Ming, A new technology of coprecipitation combined with high temperature-melting for preparing single crystal ferrite powder, Journal of magnetism and magnetic materials, 78 (1989) 73-76.
http://dx.doi.org/10.1016/0304-8853(89)90088-7

[61]   Z. Mosleh, P. Kameli, A. Poorbaferani, M. Ranjbar, H. Salamati, Structural, magnetic and microwave absorption properties of Ce-doped barium hexaferrite, Journal of Magnetism and Magnetic Materials, 397 (2016) 101-107.
http://dx.doi.org/10.1016/j.jmmm.2015.08.078

[62]   F. Khademi, A. Poorbafrani, P. Kameli, H. Salamati, Structural, magnetic and microwave properties of Eu-doped barium hexaferrite powders, Journal of superconductivity and novel magnetism, 25 (2012) 525-531.
http://dx.doi.org/10.1007/s10948-011-1323-1

[63]   R.C. Alange, P.P. Khirade, S.D. Birajdar, A.V. Humbe, K.M. Jadhav, Structural, magnetic and dielectric properties of Al-Cr co-substituted M-type barium hexaferrite nanoparticles, Journal of Molecular Structure, 1106 (2016) 460-467.
http://dx.doi.org/10.1016/j.molstruc.2015.11.004

[64]   R.K. Mudsainiyan, A.K. Jassal, M. Gupta, S.K. Chawla, Study on structural and magnetic properties of nanosized M-type Ba-hexaferrites synthesized by urea assisted citrate precursor route, Journal of Alloys and Compounds, 645 (2015) 421-428.

http://dx.doi.org/10.1016/j.jallcom.2015.04.218

[65]   I. Ali, M. Islam, M. Awan, M. Ahmad, Effects of Ga–Cr substitution on structural and magnetic properties of hexaferrite ($BaFe_{12}O_{19}$) synthesized by sol–gel auto-combustion route, Journal of Alloys and Compounds, 547 (2013) 118-125.
http://dx.doi.org/10.1016/j.jallcom.2012.08.122

[66]   G.-H. An, T.-Y. Hwang, J. Kim, J. Kim, N. Kang, S. Kim, Y.-M. Choi, Y.-H. Choa, Barium hexaferrite nanoparticles with high magnetic properties by salt-assisted ultrasonic spray pyrolysis, Journal of Alloys and Compounds, 583 (2014) 145-150.
http://dx.doi.org/10.1016/j.jallcom.2013.08.105

[67]   M.H. Kim, D.S. Jung, Y.C. Kang, J.H. Choi, Nanosized barium ferrite powders prepared by spray pyrolysis from citric acid solution, Ceramics International, 35 (2009) 1933-1937.
http://dx.doi.org/10.1016/j.ceramint.2008.10.016

[68]   D. Primc, D. Makovec, D. Lisjak, M. Drofenik, Hydrothermal synthesis of ultrafine barium hexaferrite nanoparticles and the preparation of their stable suspensions, Nanotechnology, 20 (2009) 315605.
http://dx.doi.org/10.1088/0957-4484/20/31/315605

[69]   T. Yamauchi, Y. Tsukahara, T. Sakata, H. Mori, T. Chikata, S. Katoh, Y. Wada, Barium ferrite powders prepared by microwave-induced hydrothermal reaction and magnetic property, Journal of Magnetism and Magnetic Materials, 321 (2009) 8-11.
http://dx.doi.org/10.1016/j.jmmm.2008.07.005

[70]   X. Liu, J. Wang, L.-M. Gan, S.-C. Ng, Improving the magnetic properties of hydrothermally synthesized barium ferrite, Journal of magnetism and magnetic materials, 195 (1999) 452-459.
http://dx.doi.org/10.1016/S0304-8853(99)00123-7

[71]   T. Koutzarova, S. Kolev, C. Ghelev, I. Nedkov, B. Vertruen, R. Cloots, C. Henrist, A. Zaleski, Differences in the structural and magnetic properties of nanosized barium hexaferrite powders prepared by single and double microemulsion techniques, Journal of Alloys and Compounds, 579 (2013) 174-180.
http://dx.doi.org/10.1016/j.jallcom.2013.06.049

[72]   T. Koutzarova, S. Kolev, K. Grigorov, C. Ghelev, A. Zaleski, R.E. Vandenberghe, M. Ausloos, C. Henrist, R. Cloots, I. Nedkov, Structural and magnetic properties of nanosized barium hexaferrite powders obtained by microemulsion technique, Solid State Phenomena, Trans Tech Publ, 2010, pp. 57-62.

[73]   P. Xu, X. Han, M. Wang, Synthesis and magnetic properties of $BaFe_{12}O_{19}$ hexaferrite nanoparticles by a reverse microemulsion technique, The Journal of Physical Chemistry C, 111 (2007) 5866-5870.
http://dx.doi.org/10.1021/jp068955c

[74]   Y. Liu, M.G. Drew, Y. Liu, J. Wang, M. Zhang, Preparation, characterization and magnetic properties of the doped barium hexaferrites $BaFe_{12-2x}Co_{x/2}Zn_{x/2}Sn_xO_{19}$, x = 0.0–2.0, Journal of Magnetism and Magnetic Materials, 322 (2010) 814-818.
http://dx.doi.org/10.1016/j.jmmm.2009.11.009

[75]   R.H. Arendt, The molten salt synthesis of single domain $BaFe_{12}O_{19}$ and $SrFe_{12}O_{19}$ crystals, Journal of Solid State Chemistry, 8 (1973) 339-347.
http://dx.doi.org/10.1016/S0022-4596(73)80031-3

[76]   G. Albanese, A. Deriu, Magnetic properties of Al, Ga, Sc, In substituted barium ferrites: a comparative analysis, Ceramics International, 5 (1979) 3-10.
http://dx.doi.org/10.1016/0390-5519(79)90002-4

[77]   M.H. Shams, A.S. Rozatian, M.H. Yousefi, J. Valíček, V. Šepelák, Effect of Mg^{2+} and Ti^{4+} dopants on the structural, magnetic and high-frequency ferromagnetic properties of barium hexaferrite, Journal of Magnetism and Magnetic Materials, 399 (2016) 10-18.
http://dx.doi.org/10.1016/j.jmmm.2015.08.099

[78]   Y. Han, J. Sha, L. Sun, Q. Tang, Q. Lu, H. Jin, D. Jin, H. Ge, X. Wang, Improved Magnetic Properties of Sm Combining Co Or/and Zn Substituted Barium M-Type Hexaferrites, International Journal of Modern Physics B, 26 (2012) 1250141.
http://dx.doi.org/10.1142/S021797921250141X

[79]   D.-H. Kim, Y.-K. Lee, K.-M. Kim, K.-N. Kim, S.-Y. Choi, I.-B. Shim, Synthesis of Ba-ferrite microspheres doped with Sr for thermoseeds in hyperthermia, Journal of materials science, 39 (2004) 6847-6850.
http://dx.doi.org/10.1023/B:JMSC.0000045617.92955.12

[80]   R. Pullar, A. Bhattacharya, The magnetic properties of aligned M hexa-ferrite fibres, Journal of magnetism and magnetic materials, 300 (2006) 490-499.
http://dx.doi.org/10.1016/j.jmmm.2005.06.001

[81]   S. Kulkarni, J. Shrotri, C. Deshpande, S. Date, Synthesis of chemically coprecipitated hexagonal strontium-ferrite and its characterization, Journal of materials science, 24 (1989) 3739-3744.
http://dx.doi.org/10.1007/BF02385764

[82] A. Ataie, C. Ponton, I. Harris, Heat treatment of strontium hexaferrite powder in nitrogen, hydrogen and carbon atmospheres: a novel method of changing the magnetic properties, Journal of materials science, 31 (1996) 5521-5527.
http://dx.doi.org/10.1007/BF01159326

[83] A. Ataie, I. Harris, C. Ponton, Magnetic properties of hydrothermally synthesized strontium hexaferrite as a function of synthesis conditions, Journal of materials science, 30 (1995) 1429-1433.
http://dx.doi.org/10.1007/BF00375243

[84] Z. Jin, W. Tang, J. Zhang, H. Lin, Y. Du, Magnetic properties of isotropic $SrFe_{12}O_{19}$ fine particles prepared by mechanical alloying, Journal of magnetism and magnetic materials, 182 (1998) 231-237.
http://dx.doi.org/10.1016/S0304-8853(97)00679-3

[85] A. Ataie, S. Heshmati-Manesh, Synthesis of ultra-fine particles of strontium hexaferrite by a modified co-precipitation method, Journal of the European Ceramic Society, 21 (2001) 1951-1955.
http://dx.doi.org/10.1016/S0955-2219(01)00149-2

[86] Young-Min Kang, Kyoung-Seok Moon, Magnetic properties of Ce-Mn substituted M-type Sr-hexaferrites, Ceramics International, 41 (2015) 12828-12834.
http://dx.doi.org/10.1016/j.ceramint.2015.06.119

[87] R. Palomino, A.B. Miró, F. Tenorio, F.S. De Jesús, C.C. Escobedo, S. Ammar, Sonochemical assisted synthesis of $SrFe_{12}O_{19}$ nanoparticles, Ultrasonics sonochemistry, 29 (2016) 470-475.
http://dx.doi.org/10.1016/j.ultsonch.2015.10.023

[88] A. Bolarín-Miró, F. Sánchez-De Jesús, C.A. Cortes-Escobedo, S. Diaz-De La Torre, R. Valenzuela, Synthesis of M-type $SrFe_{12}O_{19}$ by mechanosynthesis assisted by spark plasma sintering, Journal of Alloys and Compounds, 643 (2015) S226-S230.
http://dx.doi.org/10.1016/j.jallcom.2014.11.124

[89] C. Jin-Ho, H. Yang-Su, S. Seung-Wan, Preparation and magnetic properties of ultrafine $SrFe_{12}O_{19}$ particles derived from a metal citrate complex, Materials Letters, 19 (1994) 257-262.
http://dx.doi.org/10.1016/0167-577X(94)90167-8

[90] J. Ding, W. Miao, P. McCormick, R. Street, High-coercivity ferrite magnets prepared by mechanical alloying, Journal of Alloys and Compounds, 281 (1998) 32-36.
http://dx.doi.org/10.1016/S0925-8388(98)00766-X

[91] G.B. Teh, Y.C. Wong, R.D. Tilley, Effect of annealing temperature on the structural, photoluminescence and magnetic properties of sol–gel derived Magnetoplumbite-type (M-type) hexagonal strontium ferrite, Journal of Magnetism and Magnetic Materials, 323 (2011) 2318-2322.
http://dx.doi.org/10.1016/j.jmmm.2011.04.014

[92] Q. Fang, Y. Liu, P. Yin, X. Li, Magnetic properties and formation of Sr ferrite nanoparticle and Zn, Ti/Ir substituted phases, Journal of magnetism and magnetic materials, 234 (2001) 366-370.
http://dx.doi.org/10.1016/S0304-8853(01)00428-0

[93] P. Long, H. Yue-Bin, G. Cheng, L. Le-Zhong, W. Rui, H. Yun, T. Xiao-Qiang, Preparation and magnetic properties of $SrFe_{12}O_{19}$ ferrites suitable for use in self-biased LTCC circulators, Chinese Physics Letters, 32 (2015) 017502.
http://dx.doi.org/10.1088/0256-307X/32/1/017502

[94] A. Guerrero-Serrano, T. Pérez-Juache, M. Mirabal-García, J. Matutes-Aquino, S. Palomares-Sánchez, Effect of barium on the properties of lead hexaferrite, Journal of superconductivity and novel magnetism, 24 (2011) 2307-2312.
http://dx.doi.org/10.1007/s10948-011-1181-x

[95] A. Guerrero-Serrano, S. Palomares-Sánchez, M. Mirabal-García, J. Matutes-Aquino, Magneto-structural characterization of strontium substituted lead hexaferrite, Journal of superconductivity and novel magnetism, 25 (2012) 1223-1228.
http://dx.doi.org/10.1007/s10948-012-1411-x

[96] A. Guerrero, M. Mirabal-García, S. Palomares-Sánchez, J. Martínez, Effect of pb on the magnetic interactions of the M-type hexaferrites, Journal of Magnetism and Magnetic Materials, 399 (2016) 41-45.
http://dx.doi.org/10.1016/j.jmmm.2015.09.052

[97] M.N. Ashiq, R.B. Qureshi, M.A. Malana, M.F. Ehsan, Fabrication, structural, dielectric and magnetic properties of tantalum and potassium doped M-type strontium calcium hexaferrites, Journal of Alloys and Compounds, 651 (2015) 266-272.
http://dx.doi.org/10.1016/j.jallcom.2015.05.181

[98] A. Hooda, S. Sanghi, A. Agarwal, R. Dahiya, Crystal structure refinement, dielectric and magnetic properties of Ca/Pb substituted $SrFe_{12}O_{19}$ hexaferrites, Journal of Magnetism and Magnetic Materials, 387 (2015) 46-52.
http://dx.doi.org/10.1016/j.jmmm.2015.03.078

[99] S.E. Mousavi Ghahfarokhi, Z.A. Rostami, I. Kazeminezhad, Fabrication of $PbFe_{12}O_{19}$ nanoparticles and study of their structural, magnetic and dielectric properties, Journal of Magnetism and Magnetic Materials, 399 (2016) 130-142. http://dx.doi.org/10.1016/j.jmmm.2015.09.063

[100] P. Popa, E. Rezlescu, C. Doroftei, N. Rezlescu, Influence of calcium on properties of strontium and barium ferrites for magnetic media prepared by combustion, J. Optoelectron. Adv. Mater, 7 (2005) 1553-1556.

[101] P.P. Kirpichk, N.B. Voronina, A.F. Sitnikov, V.Y. Garmash, Effect of rare-earth oxides on the magnetic properties of anisotrpic barium ferrite, Izvestiya Vysshikh Uchebnykh Zavedenii, Fizika, 1 (1989) 34-39.

[102] C.-J. Li, B. Wang, J.-N. Wang, Magnetic and microwave absorbing properties of electrospun $Ba_{(1-x)}La_xFe_{12}O_{19}$ nanofibers, Journal of Magnetism and Magnetic Materials, 324 (2012) 1305-1311. http://dx.doi.org/10.1016/j.jmmm.2011.11.016

[103] X. Liu, W. Zhong, S. Yang, Z. Yu, B. Gu, Y. Du, Influences of La^{3+} substitution on the structure and magnetic properties of M-type strontium ferrites, Journal of Magnetism and Magnetic Materials, 238 (2002) 207-214. http://dx.doi.org/10.1016/S0304-8853(01)00914-3

[104] L. Peng, L. Li, R. Wang, Y. Hu, X. Tu, X. Zhong, Microwave sintered $Sr_{1-x}La_xFe_{12-x}Co_xO_{19}$ (x = 0–0.5) ferrites for use in low temperature co-fired ceramics technology, Journal of Alloys and Compounds, 656 (2016) 290-294. http://dx.doi.org/10.1016/j.jallcom.2015.08.263

[105] Y.-M. Kang, High saturation magnetization in La–Ce–Zn–doped M-type Sr-hexaferrites, Ceramics International, 41 (2015) 4354-4359. http://dx.doi.org/10.1016/j.ceramint.2014.11.125

[106] Z. Zhou, Z. Wang, X. Wang, X. Wang, J. Zhang, F. Dou, M. Jin, J. Xu, Differences in the structure and magnetic properties of $Sr_{1-x}RE_xFe_{12}O_{19}$ (RE: Pr and Dy) ferrites by microwave-assisted synthesis method, Journal of Alloys and Compounds, 610 (2014) 264-270. http://dx.doi.org/10.1016/j.jallcom.2014.04.217

[107] M. Awawdeh, I. Bsoul, S.H. Mahmood, Magnetic properties and Mössbauer spectroscopy on Ga, Al, and Cr substituted hexaferrites, Journal of Alloys and Compounds, 585 (2014) 465-473. http://dx.doi.org/10.1016/j.jallcom.2013.09.174

[108] S. Wang, J. Ding, Y. Shi, Y. Chen, High coercivity in mechanically alloyed $BaFe_{10}Al_2O_{19}$, Journal of magnetism and magnetic materials, 219 (2000) 206-212.
http://dx.doi.org/10.1016/S0304-8853(00)00450-9

[109] J. Dahal, L. Wang, S. Mishra, V. Nguyen, J. Liu, Synthesis and magnetic properties of $SrFe_{12-x-y}Al_xCo_yO_{19}$ nanocomposites prepared via autocombustion technique, Journal of Alloys and Compounds, 595 (2014) 213-220.
http://dx.doi.org/10.1016/j.jallcom.2013.12.186

[110] D. Chen, Y. Liu, Y. Li, K. Yang, H. Zhang, Microstructure and magnetic properties of Al-doped barium ferrite with sodium citrate as chelate agent, Journal of magnetism and magnetic Materials, 337 (2013) 65-69.
http://dx.doi.org/10.1016/j.jmmm.2013.02.036

[111] L. Lechevallier, J. Le Breton, J. Wang, I. Harris, Structural analysis of hydrothermally synthesized $Sr_{1-x}Sm_xFe_{12}O_{19}$ hexagonal ferrites, Journal of magnetism and magnetic materials, 269 (2004) 192-196.
http://dx.doi.org/10.1016/S0304-8853(03)00591-2

[112] B. Rai, S. Mishra, V. Nguyen, J. Liu, Synthesis and characterization of high coercivity rare-earth ion doped $Sr_{0.9}RE_{0.1}Fe_{10}Al_2O_{19}$ (RE: Y, La, Ce, Pr, Nd, Sm, and Gd), Journal of Alloys and Compounds, 550 (2013) 198-203.
http://dx.doi.org/10.1016/j.jallcom.2012.09.021

[113] D. Taft, Hexagonal ferrite isolators, Journal of Applied Physics, 35 (1964) 776-778.
http://dx.doi.org/10.1063/1.1713472

[114] S. Ounnunkad, P. Winotai, Properties of Cr-substituted M-type barium ferrites prepared by nitrate–citrate gel-autocombustion process, Journal of Magnetism and Magnetic Materials, 301 (2006) 292-300.
http://dx.doi.org/10.1016/j.jmmm.2005.07.003

[115] A.A. Nourbakhsh, M. Noorbakhsh, M. Nourbakhsh, M. Shaygan, K.J. Mackenzie, The effect of nano sized $SrFe_{12}O_{19}$ additions on the magnetic properties of chromium-doped strontium-hexaferrite ceramics, Journal of Materials Science: Materials in Electronics, 22 (2011) 1297-1302.
http://dx.doi.org/10.1007/s10854-011-0303-3

[116] S. Katlakunta, S.S. Meena, S. Srinath, M. Bououdina, R. Sandhya, K. Praveena, Improved magnetic properties of Cr^{3+} doped $SrFe_{12}O_{19}$ synthesized via microwave hydrothermal route, Materials Research Bulletin, 63 (2015) 58-66.
http://dx.doi.org/10.1016/j.materresbull.2014.11.043

[117] K. Praveena, M. Bououdina, M.P. Reddy, S. Srinath, R. Sandhya, S. Katlakunta, Structural, Magnetic, and Electrical Properties of Microwave-Sintered Cr^{3+}-Doped Sr Hexaferrites, Journal of Electronic Materials, 44 (2015) 524-531.
http://dx.doi.org/10.1007/s11664-014-3453-2

[118] A. Awadallah, S.H. Mahmood, Y. Maswadeh, I. Bsoul, A. Aloqaily, Structural and magnetic properties of Vanadium Doped M-Type Barium Hexaferrite ($BaFe_{12-x}V_xO_{19}$), IOP Conference Series: Materials Science and Engineering, IOP Publishing, 2015, pp. 012006.
http://dx.doi.org/10.1088/1757-899x/92/1/012006

[119] Puneet Sharma, R.A. Roch, S.N. Medeiros, B. Hallouche, A. Paesano Jr, Structural and magnetic studies on mechanosynthesized $BaFe_{12-x}Mn_xO_{19}$, Journal of Magnetism and Magnetic Materials, 316 (2007) 29-33.
http://dx.doi.org/10.1016/j.jmmm.2007.03.207

[120] Xin Zhang, Yuping Duan, Hongtao Guan, Shunhua Liu, B. Wen, Effect of doping MnO_2 on magnetic properties for M-type barium ferrite, Journal of Magnetism and Magnetic Materials, 311 (2007) 507-511.
http://dx.doi.org/10.1016/j.jmmm.2006.08.007

[121] A. Ghosh, A. Pasko, S.N. Kane, M. Satalkar, R. Prasad, R. Diwedi, S. Ladole, A.S. Aswar, O. G.N.P., A. Apolinario, C.T. Sousa, J.P. Araujo, F. Mazaleyrat, Influene of Mn addition on magnetic and structural properties of barium hexaferrite, Proceeding of International Conference on Recent Trends in Applied Physics and Materials Science AIP Conf. Proc. 1536, 2013, pp. 961-962.

[122] V.P. Singh, G. Kumar, R. Kotnala, J. Shah, S. Sharma, K. Daya, K.M. Batoo, M. Singh, Remarkable magnetization with ultra-low loss $BaGd_xFe_{12-x}O_{19}$ nanohexaferrites for applications up to C-band, Journal of Magnetism and Magnetic Materials, 378 (2015) 478-484.
http://dx.doi.org/10.1016/j.jmmm.2014.11.071

[123] R. Pawar, S. Desai, Q. Tamboli, S.E. Shirsath, S. Patange, Ce^{3+} incorporated structural and magnetic properties of M type barium hexaferrites, Journal of Magnetism and Magnetic Materials, 378 (2015) 59-63.
http://dx.doi.org/10.1016/j.jmmm.2014.10.166

[124] C. Tsung-Shune, D. Ming-Cheng, H. Sung-Lin, Hexaferrite particles prepared by a novel flux method with δ-FeOOH as a precursor, Materials chemistry and physics, 37 (1994) 45-51.
http://dx.doi.org/10.1016/0254-0584(94)90069-8

[125] H. Fang, Z. Yang, C. Ong, Y. Li, C. Wang, Preparation and magnetic properties of (Zn–Sn) substituted barium hexaferrite nanoparticles for magnetic recording, Journal of magnetism and magnetic materials, 187 (1998) 129-135.
http://dx.doi.org/10.1016/S0304-8853(98)00139-5

[126] G. Bottoni, D. Candolfo, A. Cecchetti, F. Masoli, Magnetic anisotropy distribution in Ba ferrite particles, Journal of magnetism and magnetic materials, 193 (1999) 326-328.
http://dx.doi.org/10.1016/S0304-8853(98)00447-8

[127] N. Kodama, H. Inoue, G. Spratt, Y. Uesaka, M. Katsumoto, Noise characteristics of barium ferrite particulate rigid disks, Journal of Applied Physics, 69 (1991) 4490-4492.
http://dx.doi.org/10.1063/1.348334

[128] N. Kodama, H. Inoue, G. Spratt, Y. Uesaka, M. Katsumoto, Media noise and interparticle interactions of barium ferrite particulate rigid disks, Journal of magnetism and magnetic materials, 116 (1992) 291-297.
http://dx.doi.org/10.1016/0304-8853(92)90175-N

[129] D. Speliotis, Distinctive characteristics of barium ferrite media, Magnetics, IEEE Transactions on, 23 (1987) 3143-3145.

[130] D. Speliotis, Barium ferrite magnetic recording media, Magnetics, IEEE Transactions on, 23 (1987) 25-28.
http://dx.doi.org/10.1109/tmag.1987.1064808

[131] D. Speliotis, Digital recording performance of Ba-ferrite media, Journal of Applied Physics, 61 (1987) 3878-3880.
http://dx.doi.org/10.1063/1.338627

[132] J. Judge, E. Speliotis, The effect of texturizing on the magnetic and recording properties of plated rigid disks, Magnetics, IEEE Transactions on, 23 (1987) 3402-3404.
http://dx.doi.org/10.1109/tmag.1987.1065392

[133] N. Nagai, N. Sugita, M. Maekawa, Formation of hexagonal, platelike Ba-ferrite particles with low temperature dependence of coercivity, Journal of magnetism and magnetic materials, 120 (1993) 33-36.
http://dx.doi.org/10.1016/0304-8853(93)91280-K

[134] S.Y. An, I.-B. Shim, C.S. Kim, Mössbauer and magnetic properties of Co–Ti substituted barium hexaferrite nanoparticles, Journal of applied physics, 91 (2002) 8465-8467.
http://dx.doi.org/10.1063/1.1452203

[135] S. Sugimoto, K. Okayama, S.-i. Kondo, H. Ota, M. Kimura, Y. Yoshida, H. Nakamura, D. Book, T. Kagotani, M. Homma, Barium M-type ferrite as an electromagnetic microwave absorber in the GHz range, Materials Transactions, JIM, 39 (1998) 1080-1083.

[136] C. Wang, L. Li, J. Zhou, X. Qi, Z. Yue, High-frequency magnetic properties of Co-Ti substituted barium ferrites prepared by modified chemical coprecipitation method, Journal of Materials Science: Materials in Electronics, 13 (2002) 713-716.
http://dx.doi.org/10.1023/A:1021560920450

[137] F. Tabatabaie, M. Fathi, A. Saatchi, A. Ghasemi, Effect of Mn–Co and Co–Ti substituted ions on doped strontium ferrites microwave absorption, Journal of Alloys and Compounds, 474 (2009) 206-209.
http://dx.doi.org/10.1016/j.jallcom.2008.06.083

[138] D. Han, Z. Yang, H. Zeng, X. Zhou, A. Morrish, Cation site preference and magnetic properties of Co-Sn-substituted Ba ferrite particles, Journal of magnetism and magnetic materials, 137 (1994) 191-196.
http://dx.doi.org/10.1016/0304-8853(94)90205-4

[139] A. Gonzalez-Angeles, G. Mendoza-Suarez, A. Gruskova, I. Toth, V. Jančárik, M. Papanova, J. Escalante-Garcí, Magnetic studies of NiSn-substituted barium hexaferrites processed by attrition milling, Journal of magnetism and magnetic materials, 270 (2004) 77-83.
http://dx.doi.org/10.1016/j.jmmm.2003.08.001

[140] A. González-Angeles, G. Mendoza-Suárez, A. Grusková, J. Sláma, J. Lipka, M. Papánová, Magnetic structure of $Sn^{2+}Ru^{4+}$-substituted barium hexaferrites prepared by mechanical alloying, Materials letters, 59 (2005) 1815-1819.
http://dx.doi.org/10.1016/j.matlet.2005.01.072

[141] A. González-Angeles, G. Mendoza-Suarez, A. Grusková, M. Papanova, J. Slama, Magnetic studies of Zn–Ti-substituted barium hexaferrites prepared by mechanical milling, Materials letters, 59 (2005) 26-31.
http://dx.doi.org/10.1016/j.matlet.2004.09.012

[142] A. González-Angeles, G. Mendoza-Suarez, A. Grusková, J. Lipka, M. Papanova, J. Slama, Effect of (Ni, Zn) Ru mixtures on magnetic properties of barium hexaferrites yielded by high-energy milling, Journal of magnetism and magnetic materials, 285 (2005) 450-455.
http://dx.doi.org/10.1016/j.jmmm.2004.08.015

[143] I. Bsoul, S.H. Mahmood, A.F. Lehlooh, A. Al-Jamel, Structural and magnetic properties of $SrFe_{12-2x}Ti_xRu_xO_{19}$, Journal of Alloys and Compounds, 551 (2013) 490-495.
http://dx.doi.org/10.1016/j.jallcom.2012.11.062

[144] G.H. Dushaq, S.H. Mahmood, I. Bsoul, H.K. Juwhari, B. Lahlouh, M.A. AlDamen, Effects of molybdenum concentration and valence state on the structural and magnetic properties of $BaFe_{11.6}Mo_xZn_{0.4-x}O_{19}$ hexaferrites, Acta Metallurgica Sinica (English Letters), 26 (2013) 509-516.
http://dx.doi.org/10.1007/s40195-013-0075-2

[145] H. Vincent, E. Brando, B. Sugg, Cationic Distribution in Relation to the Magnetic Properties of New M-Hexaferrites with Planar Magnetic Anisotropy $BaFe_{12-2x}Ir_xMe_xO_{19}$ (Me= Co, Zn, x≈ 0.85 and x≈ 0.50), Journal of Solid State Chemistry, 120 (1995) 17-22.
http://dx.doi.org/10.1006/jssc.1995.1369

[146] B. Sugg, H. Vincent, Magnetic properties of new M-type hexaferrites $BaFe_{12-2x}Ir_xCo_xO_{19}$, Journal of magnetism and magnetic materials, 139 (1995) 364-370.
http://dx.doi.org/10.1016/0304-8853(95)90016-0

[147] S.H. Mahmood, G.H. Dushaq, I. Bsoul, M. Awawdeh, H.K. Juwhari, B.I. Lahlouh, M.A. AlDamen, Magnetic Properties and Hyperfine Interactions in M-Type $BaFe_{12-2x}Mo_xZn_xO_{19}$ Hexaferrites, Journal of Applied Mathematics and Physics, 2 (2014) 77-87.
http://dx.doi.org/10.4236/jamp.2014.25011

[148] M.V. Rane, D. Bahadur, S. Mandal, M. Patni, Characterization of $BaFe_{12-2x}Co_xZr_xO_{19}$ ($0≤ x≤ 0.5$) synthesised by citrate gel precursor route, Journal of magnetism and magnetic materials, 153 (1996) L1-L4.
http://dx.doi.org/10.1016/0304-8853(95)00305-3

[149] P. Wartewig, M. Krause, P. Esquinazi, S. Rösler, R. Sonntag, Magnetic properties of Zn-and Ti-substituted barium hexaferrite, Journal of magnetism and magnetic materials, 192 (1999) 83-99.
http://dx.doi.org/10.1016/S0304-8853(98)00382-5

[150] F. Wei, H. Fang, C. Ong, C. Wang, Z. Yang, Magnetic properties of $BaFe_{12-2x}Zn_xZr_xO_{19}$ particles, Journal of Applied Physics, 87 (2000) 8636-8639.
http://dx.doi.org/10.1063/1.373589

[151] G. Mendoza-Suarez, L. Rivas-Vazquez, J. Corral-Huacuz, A. Fuentes, J. Escalante-Garcí, Magnetic properties and microstructure of $BaFe_{11.6-2x}Ti_xM_xO_{19}$ (M= Co, Zn, Sn) compounds, Physica B: Condensed Matter, 339 (2003) 110-118.

http://dx.doi.org/10.1016/j.physb.2003.08.120

[152] A.M. Alsmadi, I. Bsoul, S.H. Mahmood, G. Alnawashi, K. Prokeš, K. Siemensmeyer, B. Klemke, H. Nakotte, Magnetic study of M-type doped barium hexaferrite nanocrystalline particles, Journal of Applied Physics, 114 (2013) 243910.
http://dx.doi.org/10.1063/1.4858383

[153] Z. Yang, C. Wang, X. Li, H. Zeng, (Zn, Ni, Ti) substituted barium ferrite particles with improved temperature coefficient of coercivity, Materials Science and Engineering: B, 90 (2002) 142-145.
http://dx.doi.org/10.1016/S0921-5107(01)00925-4

[154] S. Pignard, H. Vincent, E. Flavin, F. Boust, Magnetic and electromagnetic properties of RuZn and RuCo substituted $BaFe_{12}O_{19}$, Journal of magnetism and magnetic materials, 260 (2003) 437-446.
http://dx.doi.org/10.1016/S0304-8853(02)01387-2

[155] X. Batlle, X. Obradors, J. Rodriguez-Carvajal, M. Pernet, M. Cabanas, M. Vallet, Cation distribution and intrinsic magnetic properties of Co-Ti-doped M-type barium ferrite, Journal of applied physics, 70 (1991) 1614-1623.
http://dx.doi.org/10.1063/1.349526

[156] X. Zhou, A. Morrish, Z. Li, Y. Hong, Site preference for Co^{2+} and Ti^{4+} in Co-Ti substituted barium ferrite, Magnetics, IEEE Transactions on, 27 (1991) 4654-4656.

[157] A. Gruskova, J. Slama, M. Michalikova, J. Lipka, I. Toth, P. Kaboš, Preparation of substituted barium ferrite powders, Journal of magnetism and magnetic materials, 101 (1991) 227-229.
http://dx.doi.org/10.1016/0304-8853(91)90738-V

[158] A. Morrish, X. Zhou, Z. Yang, H.-X. Zeng, Substituted barium ferrites; sources of anisotropy, Hyperfine Interactions, 90 (1994) 365-369.
http://dx.doi.org/10.1007/BF02069140

[159] Z. Šimša, S. Lego, R. Gerber, E. Pollert, Cation distribution in Co-Ti-substituted barium hexaferrites: a consistent model, Journal of magnetism and magnetic materials, 140 (1995) 2103-2104.
http://dx.doi.org/10.1016/0304-8853(94)01393-4

[160] G. Bottoni, Magnetization stability and interactions in particulate recording media, Materials chemistry and physics, 42 (1995) 45-50.
http://dx.doi.org/10.1016/0254-0584(95)01551-5

[161] K. Kakizaki, N. Hiratsuka, T. Namikawa, Fine structure of acicular $BaCo_xTi_xFe_{12-2x}O_{19}$ particles and their magnetic properties, Journal of magnetism and magnetic materials, 176 (1997) 36-40.
http://dx.doi.org/10.1016/S0304-8853(97)00634-3

[162] Y. Li, R. Liu, Z. Zhang, C. Xiong, Synthesis and characterization of nanocrystalline $BaFe_{9.6}Co_{0.8}Ti_{0.8}M_{0.8}O_{19}$ particles, Materials chemistry and physics, 64 (2000) 256-259.
http://dx.doi.org/10.1016/S0254-0584(99)00218-7

[163] G. Mendoza-Suarez, J. Corral-Huacuz, M. Contreras-García, H. Juarez-Medina, Magnetic properties of $BaFe_{11.6-2x}Co_xTi_xO_{19}$ particles produced by sol–gel and spray-drying, Journal of magnetism and magnetic materials, 234 (2001) 73-79.
http://dx.doi.org/10.1016/S0304-8853(01)00286-4

[164] M. Kuznetsov, Q. Pankhurst, I. Parkin, Novel SHS routes to CoTi-doped M-type ferrites, Journal of Materials Science: Materials in Electronics, 12 (2001) 533-536.
http://dx.doi.org/10.1023/A:1012405610723

[165] A. Gruskova, J. Slama, R. Dosoudil, D. Kevicka, V. Jančárik, I. Toth, Influence of Co–Ti substitution on coercivity in Ba ferrites, Journal of magnetism and Magnetic Materials, 242 (2002) 423-425.
http://dx.doi.org/10.1016/S0304-8853(01)01139-8

[166] C. Wang, X. Qi, L. Li, J. Zhou, X. Wang, Z. Yue, High-frequency magnetic properties of low-temperature sintered Co-Ti substituted barium ferrites, Materials Science and Engineering: B, 99 (2003) 270-273.
http://dx.doi.org/10.1016/S0921-5107(02)00521-4

[167] Z. Haijun, L. Zhichao, M. Chenliang, Y. Xi, Z. Liangying, W. Mingzhong, Preparation and microwave properties of Co-and Ti-doped barium ferrite by citrate sol–gel process, Materials chemistry and physics, 80 (2003) 129-134.
http://dx.doi.org/10.1016/S0254-0584(02)00457-1

[168] R. Lima, M.S. Pinho, M.L. Gregori, R.R. Nunes, T. Ogasawara, Effect of double substituted *m*-barium hexaferrites on microwave absorption properties, Materials Science-Poland, 22 (2004) 245-252.

[169] L. Jia, H. Zhang, S. Yin, F. Bai, B. Liu, Q. Wen, J. Shen, Synthesis and magnetic properties of Co–Ti–Bi codoped M-type barium ferrite, Journal of Applied Physics, 109 (2011) 07E317.

[170] A. Alsmadi, I. Bsoul, S. Mahmood, G. Alnawashi, F. Al-Dweri, Y. Maswadeh, U. Welp, Magnetic study of M-type Ru-Ti doped strontium hexaferrite nanocrystalline particles, Journal of Alloys and Compounds, 648 (2015) 419-427. *http://dx.doi.org/10.1016/j.jallcom.2015.06.274*

[171] O. Kubo, E. Ogawa, Barium ferrite particles for high density magnetic recording, Journal of magnetism and magnetic materials, 134 (1994) 376-381. *http://dx.doi.org/10.1016/0304-8853(94)00147-2*

[172] M.N. Ashiq, M.J. Iqbal, I.H. Gul, Structural, magnetic and dielectric properties of Zr–Cd substituted strontium hexaferrite ($SrFe_{12}O_{19}$) nanoparticles, Journal of Alloys and Compounds, 487 (2009) 341-345. *http://dx.doi.org/10.1016/j.jallcom.2009.07.140*

[173] S.H. Mahmood, A. Awadallah, Y. Maswadeh, I. Bsoul, Structural and magnetic properties of Cu-V substituted M-type barium hexaferrites, IOP Conference Series: Materials Science and Engineering, IOP Publishing, 2015, pp. 012008. *http://dx.doi.org/10.1088/1757-899x/92/1/012008*

[174] Y. Hong, C. Ho, H.Y. Hsu, C. Liu, Synthesis of nanocrystalline $Ba(MnTi)_xFe_{12-2x}O_{19}$ powders by the sol–gel combustion method in citrate acid–metal nitrates system ($x = 0, 0.5, 1.0, 1.5, 2.0$), Journal of magnetism and magnetic materials, 279 (2004) 401-410. *http://dx.doi.org/10.1016/j.jmmm.2004.02.008*

[175] A. Gruskova, J. Lipka, M. Papanova, D. Kevicka, A. Gonzalez, G. Mendoza, I. Toth, J. Slama, Mössbauer study of microstructure and magnetic properties (Co, Ni)–Zr substituted Ba ferrite particles, Hyperfine interactions, 156 (2004) 187-194. *http://dx.doi.org/10.1023/B:HYPE.0000043223.50583.4a*

[176] J. Zhou, H. Ma, M. Zhong, G. Xu, Z. Yue, Z. He, Influence of Co–Zr substitution on coercivity in Ba ferrites, Journal of magnetism and magnetic materials, 305 (2006) 467-469. *http://dx.doi.org/10.1016/j.jmmm.2006.02.085*

[177] X. Gao, Y. Du, X. Liu, P. Xu, X. Han, Synthesis and characterization of Co–Sn substituted barium ferrite particles by a reverse microemulsion technique, Materials Research Bulletin, 46 (2011) 643-648. *http://dx.doi.org/10.1016/j.materresbull.2011.02.002*

CHAPTER 5

Permanent Magnet Applications

S.H. Mahmood

Physics Department, The University of Jordan, Amman, Jordan

s.mahmood@ju.edu.jo

Abstract

Permanent magnets are used in a wide range of applications, providing convenience, cost effectiveness, device miniaturization, as well as extending the operational capacities beyond those of conventional devices. In this chapter, some of the permanent magnet applications where hard ferrites are practically in use are addressed. In particular, the main energy conversion devices such as actuators and transducers, as well as some of the important microwave devices utilizing ferrite magnets are discussed.

Keywords

Motor; Transducer; Microwave Absorption, Passive Microwave Devices

Contents

1. INTRODUCTION

It is difficult nowadays to imagine our modern life without magnets. Machines and devices employing permanent magnet components have become essential for running our day-to-day routines, and providing unimaginable convenience, efficiency, as well as the capability of accomplishing jobs. Permanent magnets are found everywhere around. In house hold appliances and consumer electronics, they are found in juicers, mixers, refrigerators, washing machines, can openers, small electric tools, audio-video systems, televisions, clocks, toys, and many more. In office supplies and data processing systems, they are found in printers, computers, scanners, disc drives and actuators, paper shredders, and more. Magnets are also essential components in telecommunication applications, such as in telephone ringers and vibrators, speakers, microphones, antennas, switches and relays, filters, phase shifters, and circulators. Also, magnets are widely used in the industrial sector in automation and control processes. In a modern car, for example, many magnets are used in small motors for windows, mirrors, seat and roof actuators, wipers, pumps, fans, and in breaking systems, loudspeakers, and alternators. In addition, magnets have been used in many other applications in industry and instrumentation, aerospace, defense, and health sector [1-4].

2. CLASSIFICATIONS OF PERMANENT MAGNET APPLICATIONS

Contrary to electromagnets, permanent magnets are used to provide magnetic flux in a usable space or air gap in a magnetic circuit without spending energy. Depending on the type of application, the flux density in the usable space can be uniform or non-uniform, and the field strength can be steady in time or time-dependent [5]. In view of the wide range of applications of magnetic materials, it is a difficult task to classify these applications in a single standard way. Classifications based on the function of the device and field of application, physical principle being used in the application [4], or the type of material and its efficiency may be adopted. Classification based on the type of magnetic material and their magnetic properties was addressed in chapter 2 of this book, as well as in other sources [4, 6]. In some other cases, for example, applications were classified into static and dynamic devices [7].

Based on function, the *electrical-to-mechanical* energy conversion devices dominate the world market of permanent magnet applications. The basic principle of operation of these types of applications is the induced mechanical motion of a current-carrying coil placed in a magnetic field by the force exerted on the magnetic moment of the coil by the external field. The two main applications in this category are loudspeakers and permanent magnet motors. The *mechanical-to-electrical* energy conversion devices, on the other hand, work on the principle of converting mechanical motion, such as the vibration or rotation of a coil in a magnetic field, into electric current in the coil. Renewable energy generation based on wind power is a growing field of application under this category [6]. Further, devices made for *force applications* include magnetic chucks, latches, and magnetic separators [8]. In addition, materials for *non-microwave applications*, *microwave applications* [9], *magnetic recording applications* [10], and *magnetic refrigeration* [6], were addressed as separate classes.

Under the special category of dynamical, electrical-to-mechanical energy conversion devices, loudspeakers account for the largest consumption of permanent magnets in a single application domain [8]. Under the same category, permanent magnet motors are widely used for countless applications [2, 11, 12], and account for a large market share of permanent magnet devices, estimated at US$91.75 billion in 2015, and expected to rise at an annual rate of 6.38% during the next five years [13]. In addition, small devices such as magnetic latches for home and office use, as well as electric toys, are produced in large quantities every year. All of such applications account for a large portion of the world market of permanent magnet devices. Accordingly, abundant and low cost raw materials for permanent magnet production are ideal for these types of applications, which could be translated into large scale savings in production and sales prices of the final products. Magnetically hard hexaferrites, such as barium or strontium ferrites, could be the best candidates for offering such opportunities. These magnets are the most cost-effective components in speakers, as well as in small dc motors used in blowers, fans, pumps, wipers, electric toys and a variety of applications for home and office uses.

The remaining of this chapter will be dedicated to the discussion of a selection of main applications of permanent magnets. In view of the large diversity of permanent magnet applications, the following discussion is not intended to be extensive, neither in scope, nor in depth. For more details, the reader is referred to references cited above.

3.   SELECTED PERMANENT MAGNET APPLICATIONS

In this section, some of the permanent magnet applications are briefly discussed, without reference to any classification system. The discussion is concerned mainly with the basic principles of operation and function of device utilizing hexaferrite materials.

3.1   LOUDSPEAKERS

The dynamic speaker is a *transducer* constructed to convert a variable electrical audio signal into the corresponding sound. The basic components of a dynamic speaker, are illustrated in Fig. 1.

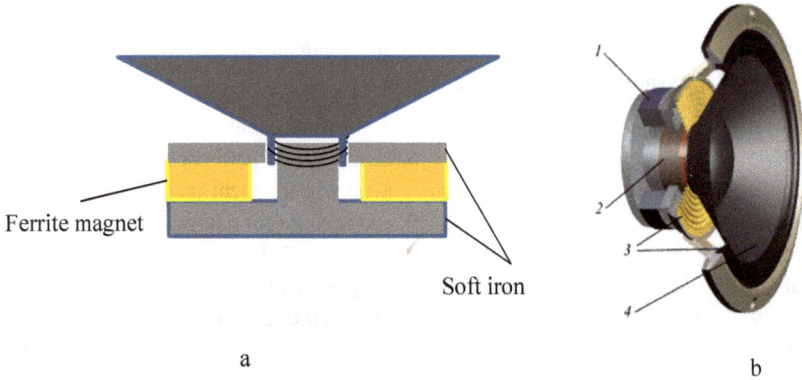

Fig. 1: (a) Schematic diagram of a dynamic speaker utilizing ferrite ring magnet on a soft iron base with a cylindrical post at center, and a soft iron ring on top. (b) a photograph of a speaker showing its main components: (1) the permanent magnet, (2) the voice coil, (3) suspension system, and (4) the diaphragm.

A lightweight diaphragm, usually in the shape of a cone or a dome, carries a coil (the voice coil) wound around its base. The voice coil fits into an annular gap of the magnetic circuit, and is free to move axially in the gap. The speaker is usually connected to a rigid *frame*, and free axial motion of the cone/voice coil assembly is maintained by utilizing a suspension system consisting of two parts, the *spider*, and the *surround*. The spider ensures axial movement of the voice coil through the cylindrical magnetic gap, while the surround is designed to center the coil/cone assembly, allowing its free piston-like motion aligned with the magnetic gap.

The variable electrical audio signal passing through the voice coil, which should be amplified for better performance of the speaker, creates a magnetic field which varies with the current in the voice coil. The electromagnetic interaction between the magnetic system and the variable coil field, governed by Faraday's low of induction, generate a mechanical force that causes the coil and the attached cone to vibrate axially in the air gap, thus reproducing the sound corresponding to the electrical signal.

In special designs of *loudspeakers*, several speakers are attached and arranged in the interior of a specially designed box made of wood, plastic, or other materials, which plays an important role in the quality of the produced sound. Some of the speakers used in such designs are constructed to reproduce low-frequency sounds (*woofers*), some are capable of reproducing *mid-range* frequencies, while others are made for the reproduction of high-frequency sound signals (*tweeters*). The whole assembly, called *loudspeaker*, is capable of reproducing high quality sound in the entire audible frequency range.

3.2 PERMANENT MAGNET MOTORS

A motor is an *actuator* constructed to produce rotary motion via the interaction between current carrying coils and a magnetic field. The basic structure of a motor is composed of a *rotor* placed within the *stator*, with a gap in between to allow free rotation of the rotor inside the stator. The field necessary to rotate the rotor is provided by the stator, which is the stationary part of the motor, and could be constructed by either a soft magnetic material core wound usually with copper windings, or permanent magnets. To reduce energy losses in wound-field stators, the soft magnetic core is made of thin sheets, called *laminations*. In small motors, the wound-field stator is replaced by a permanent magnet, which provides the necessary field without expenditure of energy.

In a simple small motor, the rotor consists of a cylindrical steel core fixed on an axial shaft and slotted into three evenly spaced sections (Fig. 2). Each of the three sections is wound with copper windings, which when a steady electric current passes through, magnetic moments develop around the cylinder spaced at 120° from one another. The interaction of the magnetic moments of the windings with the magnetic field of the stator results in a torque which rotates the rotor cylinder and its shaft. The shaft can then be fitted with a special mechanism, such as a gear system, to deliver the motion to the required parts of a body.

Although the construction discussed above seems to be an oversimplification, it serves to clarify the basic components and principles of operation of a motor. This is, however, suitable only for driving small loads, and for purposes of device miniaturization and energy savings. Large DC motors with wound-filed stators and special designs of rotors cannot be replaced in heavy uses such as propelling automobiles, boats, or elevators. The advancements achieved in the field of power electronics made it possible to replace the DC motors by less costly AC motors.

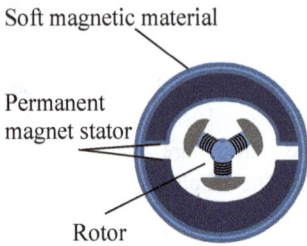

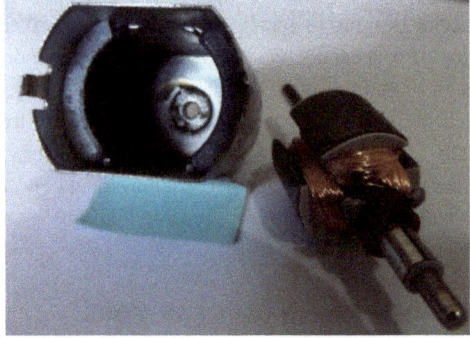

(a)   (b)

Fig. 2: (a) Basic design of a 3-phase permanent magnet motor, and (b) a photograph of the rotor, permanent magnet stator, and soft iron casing of a small motor from a toy.

3.3   MICROWAVE APPLICATIONS

Hexaferrites were recognized as good candidates for a variety of microwave applications due to their high anisotropy field and low eddy current losses [1]. These applications can be subdivided into two main categories. The first includes electromagnetic (EM) wave absorbers and interference suppressors (EMI). The second includes *non-reciprocal microwave devices* operating at ultrahigh frequencies such as *isolators*, *circulators*, *filters*, and *phase shifters* [9, 14, 15].

The above mentioned microwave devices are used to control microwave propagation in a given direction in a waveguide by use of a ferrite magnet having a controllable dynamic permeability. When an EM wave is inserted at one end of the ferrite magnet, its permeability is modified via interaction with the EM filed, and becomes critically dependent on the polarization of the EM radiation with respect to the magnetization direction. This property is essential to facilitate non-reciprocal device functioning, allowing wave propagation along one direction in a waveguide, and attenuating waves propagating in the opposite direction. Also, the dynamic permeability of the ferrite is dependent on the magnetization, as well as the intensity of the applied static *bias* magnetic field, which allows for tuning the ferrite component for operation at the desired microwave frequency. A *phase shifter* uses, in addition, the property of *Faraday rotation* of the plane of polarization of the microwave as it passes through the ferrite magnet. This is an important element in oscillators and phased-array antennas.

In our forthcoming discussion, we will provide a brief presentation of the basic principles and operation of microwave devices. Our discussion is neither intended to be comprehensive, nor to provide details of the theory and technical operation which can be found in many books and reviews provided in references [1, 9, 15, 16].

To understand the basis of this operation, we consider a static bias field applied to a magnet in the shape of an *ellipsoid of revolution* along the z-axis. Upon insertion of a circularly polarized microwave whose *rf field* rotates in the x-y plane, the magnetization vector interacts with the rf field, developing a small dynamic component in the x-y plane, and leading to precession of the magnetization vector around the static bias field. The static bias field intensity determines the *natural frequency* of precession (Larmor frequency). The dynamic permeability arising from the circularly polarized rf field is then expressed in terms of two opposite helicities corresponding to the two counter-rotating modes of the rf field. The positive helicity permeability, μ_+, is associated with rotation in the same direction of the precessional motion, whereas the negative helicity permeability, μ_-, is associated with rotation of the rf field in an opposite sense. The positive helicity permeability has the characteristic of *resonance*, which is characterized by the *resonance field* and *line width*. When the rf frequency of the incident radiation matches the natural frequency of the material (at resonance), the imaginary part of the permeability (μ_+'') exhibits a maximum, and the electromagnetic energy is fully absorbed by the spin dynamics. The precession is subsequently damped by the various damping mechanisms, transforming the absorbed energy into heat in the crystal. Thus, propagation of microwaves in the corresponding direction is highly attenuated. On the other hand, the negative helicity permeability does not exhibit resonance due to the asymmetry of the permeability tensor (*Polder tensor*) of the ferrite. Consequently, in the opposite direction of propagation, where the sense of rotation of the rf field is reversed, the imaginary part of the permeability (μ_-'') remains close to zero at any frequency, allowing wave propagation in the direction corresponding to the negative helicity. Accordingly, selecting the appropriate applied bias field intensity for matching the precessional frequency with the rf frequency is the key criterion for non-reciprocal device operation.

3.3.1 MICROWAVE ABSORBERS

Microwave absorption applications utilize ferrites with high absorption capacity for electromagnetic radiation. Ferrite beads and coatings are practical solutions used to eliminate or reduce EMI effects by absorption of unwanted high-frequency EM noise in the operational environments of electrical wires and circuits. Also, EM absorbing materials are used for defense and military applications in avoiding being detected by a hostile radar system. When a metal surface, such as that of a tank or a jet fighter, is coated with a layer of the absorbing material (metal-backed single-layer absorber), absorption of radar radiation falling onto the surface is maximized by two conditions: the first is to allow the incident EM radiation to penetrate through the absorber layer without reflection, and the second is to allow maximum radiation absorption in the material. The first condition is met by impedance matching between the absorber and free space dictated by the condition $\sqrt{\mu/\varepsilon} = 1$, where μ and ε are respectively the magnetic permeability and dielectric permittivity of the absorbing material at the frequency of the incident EM radiation. The second condition is satisfied if the absorbing material has a large imaginary part of the permittivity or the permeability at the incident radiation frequency. Practically, the real part of the permeability could be small at microwave frequencies compared to the real part of the permittivity, resulting in a small penetration depth into the material, and a substantial reflection of the incident radiation. Accordingly, for efficient absorption, impedance matching should be improved by increasing the permeability and decreasing the permittivity at the operational frequency, in addition to maximizing the imaginary parts. This can be done by modifying the properties of the material by adopting a suitable synthesis route and using suitable metal cation substitutions [17-24].

3.3.2 ISOLATORS

An *isolator* is a two-port EM waveguide section with a low insertion loss and high isolation, and thus allows propagation of the EM radiation from input to output without significant attenuation, while propagation in the opposite direction from output to input is highly attenuated. This is simply the *resonance isolator*, which is used for different applications, such as in delivering microwave radiation from a microwave generator to a load. The isolator would insure full transmission of the microwave power to the load, and inhibit back flow of reflected radiation to the generator. Also, it is used to reduce effects of frequency pulling in an oscillator due to reflections from a mismatched load. Further, it is used in high-gain traveling wave tube amplifiers [9]. There are different kinds of isolators in practical applications, some of which are discussed in the forthcoming section.

3.3.3 CIRCULATORS

A circulator is an *n*-port microwave passive component used in high frequency circuits for routing the outgoing or incoming microwave signals, and isolating the different components of the circuit. If the ports are numbered sequentially, a signal inserted in port 1 emerges at port 2, and a signal inserted through port 2 emerges at port 3, and so on. The operation of a 4-port circulator is based of Faraday rotation of the microwave field in the magnetized ferrite. The basis of operation of a 3-port circulator, on the other hand, is wave cancellation of propagating modes in two different paths in the field of the ferrite magnet [9]. In a system adopting a single antenna in the transmit/receive module, such as the radar system and the mobile communication systems, a 3-port Y-junction circulator is used, whose basic principle of operation is illustrated in Fig. 3. If the third port of the circulator is terminated by a matched load, it becomes an isolator (the *junction isolator*). Other types of isolators are discussed elsewhere

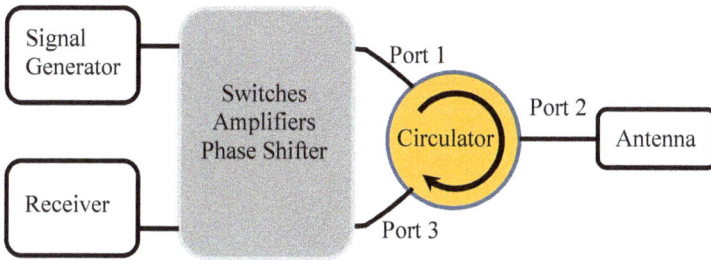

Fig. 3: A Block diagram of a microwave transmit/receive module.

Hexaferrites with their superior magnetic properties to those of the traditional cubic ferrites and garnets offer unique opportunities for device miniaturization and extending the frequency range of operation. The high magnetocrystalline anisotropy in these ferrites facilitates the occurrence of a stable magnetized state after removal of the applied field. This provides the opportunity to use a *self-biased* hexaferrite magnet, thus reducing cost and volume of the device. Further, the ability to tune the magnetic properties of BaM or SrM hexaferrites by suitable cationic substitutions also provides the opportunity to extend the operational frequency of microwave devices. Specifically, Taft [25] had prepared a variety of Al-substituted hexaferrites with anisotropy fields in the range 15 – 34 kOe, to cover the frequency range 50 – 90 GHz for use in isolators without the need for an external bias field. Al substituted SrM hexaferrites were reported to cover the frequency range of 60 – 90 GHz, while Ba-Ni$_2$ W-type hexaferrites with different levels of Al substitution were found to be suitable for the frequency range 50 – 60 GHz. Consequently, waveguide isolators operating in the bands 50 – 75 GHz and 60 – 90 GHz were developed using these materials.

The requirement of a high remanent magnetization, however, requires high coercivity, which can be attained in small grained polycrystalline hexaferrites, while the requirement of a small ferromagnetic resonance linewidth for efficient operation of the microwave device is usually optimized for a single crystal. Accordingly, a magnetically textured hexaferrite sample with properties approximating those of a single crystal could be used. At this point, it is worth mentioning that improving the microwave device by using hexaferrite component is not limited to circulators, but can also be adopted in the production of other passive microwave devices.

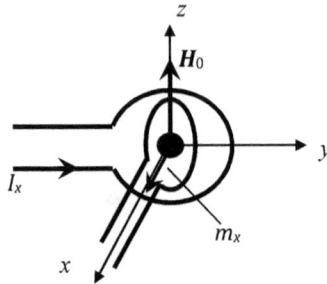

Fig. 4: Schematic diagram illustrating the basic structure of a microwave filter.

3.3.4 FILTERS

A microwave filter is a device used in circuits for transmitting microwaves with frequency falling within a desired *pass band* without significant losses, while attenuating waves with frequencies outside the pass band. The basic idea of the operation of the filter relies on using two coils arranged in such a way that their axes are perpendicular (coinciding with the x and y axes, for example) and intersect at the centers of the coils, where a ferrite sphere is situated (see Fig. 4). Normally, a DC bias (H_0) is applied to the ferrite, which is usually an yttrium iron garnet, in the z-direction. If a current (I_x) is passed in the x-coil, its moment (m_x) precesses about the applied field, producing a magnetic moment component in the y direction, which precesses at the ferromagnetic resonance frequency of the ferrite. Accordingly, an oscillating electromotive force (and current) develops in the y coil. Frequencies outside the resonance region are therefore not allowed to pass. The quality factor (Q-factor) determines the pass band width, and this factor can be controlled by the device design, the material used, and details of the coupling circuit. Yttrium iron garnets (YIG) are used in telecommunication to cover the frequency range 500 MHz – 26.5 GHz with a suitable choice of the bias field. Self-biased hexaferrite components can be useful in extending the operating frequency to higher bands, and reducing the size and power consumption as discussed previously.

REFERENCES

[1] Ü. Özgür, Y. Alivov, H. Morkoç, Microwave ferrites, part 1: fundamental properties, Journal of Materials Science: Materials in Electronics, 20 (2009) 789-834.
http://dx.doi.org/10.1007/s10854-009-9923-2

[2] J.F. Gieras, Permanent magnet motor technology: design and applications, CRC press 2002.

[3] S.-M. Lu, A review of high-efficiency motors: Specification, policy, and technology, Renewable and Sustainable Energy Reviews, 59 (2016) 1-12.
http://dx.doi.org/10.1016/j.rser.2015.12.360

[4] K.J. Strnat, Modern permanent magnets for applications in electro-technology, Proceedings of the IEEE, 78 (1990) 923-946.
http://dx.doi.org/10.1109/5.56908

[5] J. Coey, Permanent magnet applications, Journal of Magnetism and Magnetic Materials, 248 (2002) 441-456.
http://dx.doi.org/10.1016/S0304-8853(02)00335-9

[6] O. Gutfleisch, M.A. Willard, E. Brück, C.H. Chen, S. Sankar, J.P. Liu, Magnetic materials and devices for the 21st century: stronger, lighter, and more energy efficient, Advanced materials, 23 (2011) 821-842.
http://dx.doi.org/10.1002/adma.201002180

[7] H. Kirchmayr, Permanent magnets and hard magnetic materials, Journal of Physics D: Applied Physics, 29 (1996) 2763-2778.
http://dx.doi.org/10.1088/0022-3727/29/11/007

[8] B.D. Cullity, C.D. Graham, Introduction to magnetic materials, John Wiley & Sons2011.

[9] Ü. Özgür, Y. Alivov, H. Morkoç, Microwave ferrites, part 2: passive components and electrical tuning, Journal of Materials Science: Materials in Electronics, 20 (2009) 911-952.
http://dx.doi.org/10.1007/s10854-009-9924-1

[10] E.P. Wohlfarth, Ferromagnetic materials: a handbook on the properties of magnetically ordered substances, Elsevier1980.

[11]   M. Kramer, R. McCallum, I. Anderson, S. Constantinides, Prospects for non-rare earth permanent magnets for traction motors and generators, JOM, 64 (2012) 752-763.
http://dx.doi.org/10.1007/s11837-012-0351-z

[12]   M. Olszewski, Final Report on Assessment of Motor Technologies for Traction Drives of Hybrid and Electric Vehicles, , Oak Ridge National Laboratory, Washington D.C., 2011.

[13]   http://www.marketsandmarkets.com/Market-Reports/electric-motor-market-alternative-fuel-vihicles-717.html.

[14]   E. Schloemann, Advances in ferrite microwave materials and devices, Journal of magnetism and Magnetic Materials, 209 (2000) 15-20.
http://dx.doi.org/10.1016/S0304-8853(99)00635-6

[15]   V.G. Harris, A. Geiler, Y. Chen, S.D. Yoon, M. Wu, A. Yang, Z. Chen, P. He, P.V. Parimi, X. Zuo, Recent advances in processing and applications of microwave ferrites, Journal of Magnetism and Magnetic Materials, 321 (2009) 2035-2047.

[16]   J. Nicolas, Microwave ferrites, in: E.P. Wohlfarth (Ed.) Ferromagnetic Materials, North-Holland Publishing Company, New York, 1980, pp. 243-296.

[17]   S. Choopani, N. Keyhan, A. Ghasemi, A. Sharbati, R.S. Alam, Structural, magnetic and microwave absorption characteristics of $BaCo_xMn_xTi_{2x}Fe_{12-4x}O_{19}$, Materials Chemistry and Physics, 113 (2009) 717-720.
http://dx.doi.org/10.1016/j.matchemphys.2008.07.130

[18]   Y. Yang, B. Zhang, W. Xu, Y. Shi, N. Zhou, H. Lu, Microwave absorption studies of W-hexaferrite prepared by co-precipitation/mechanical milling, Journal of magnetism and magnetic materials, 265 (2003) 119-122.
http://dx.doi.org/10.1016/S0304-8853(03)00237-3

[19]   P. Meng, K. Xiong, L. Wang, S. Li, Y. Cheng, G. Xu, Tunable complex permeability and enhanced microwave absorption properties of $BaNi_xCo_{1-x}TiFe_{10}O_{19}$, Journal of Alloys and Compounds, 628 (2015) 75-80.
http://dx.doi.org/10.1016/j.jallcom.2014.10.163

[20]   Z. Mosleh, P. Kameli, A. Poorbaferani, M. Ranjbar, H. Salamati, Structural, magnetic and microwave absorption properties of Ce-doped barium hexaferrite, Journal of Magnetism and Magnetic Materials, 397 (2016) 101-107.
http://dx.doi.org/10.1016/j.jmmm.2015.08.078

[21] E. Kiani, A.S. Rozatian, M.H. Yousefi, Structural, magnetic and microwave absorption properties of $SrFe_{12-2x}(Mn_{0.5}Cd_{0.5}Zr)_xO_{19}$ ferrite, Journal of Magnetism and Magnetic Materials, 361 (2014) 25-29.
http://dx.doi.org/10.1016/j.jmmm.2014.02.042

[22] M.H. Shams, S.M.A. Salehi, A. Ghasemi, Electromagnetic wave absorption characteristics of Mg–Ti substituted Ba-hexaferrite, Materials letters, 62 (2008) 1731-1733.
http://dx.doi.org/10.1016/j.matlet.2007.09.073

[23] M.H. Shams, A.S.H. Rozatian, M.H. Yousefi, Enhancing static and dynamic magnetic properties of Mg-Zn doped Co_2Y-type hexaferrite as broadband micowave absorbing material, Journal of Optoelectronics and Advanced Materials, 17 (2015) 614-622.

[24] R.S. Alam, M. Moradi, H. Nikmanesh, J. Ventura, M. Rostami, Magnetic and microwave absorption properties of $BaMg_{x/2}Mn_{x/2}Co_xTi_{2x}Fe_{12-4x}O_{19}$ hexaferrite nanoparticles, Journal of Magnetism and Magnetic Materials, 402 (2016) 20-27.
http://dx.doi.org/10.1016/j.jmmm.2015.11.038

[25] D. Taft, Hexagonal ferrite isolators, Journal of Applied Physics, 35 (1964) 776-778.
http://dx.doi.org/10.1063/1.1713472

CHAPTER 6

Magnetic Recording

I.O. Abu Aljarayesh

Physics department, Yarmouk University, Irbid, Jordan

ijaraysh@yu.edu.jo

Abstract

In this chapter, the static magnetic properties of single domain magnetic particles, as well as the remanence thermal stability of magnetic recording materials, which are essential to the magnetic recording performance were reviewed. The main characteristics of the recording heads were also discussed. Finally, the heat assisted magnetic recording improvements were briefly addressed.

Keywords

Hysteresis; Remanence, Coercivity; Thermal Stability; Recording processes

Contents

1. INTRODUCTION

The technology of magnetic information storage continues to expand in terms of large capacity, high areal density, high transfer rate, reliability, and cost-effectiveness. Also, the improvements on the existing recording materials, and fabricating new ones, and designing and finding new recording heads, are of considerable importance to the researchers in the field [1-9]. It is relevant to mention that other properties of the recording media, like coating, lubricant, mechanical stability, design corrosion, are as equally important as the intrinsic magnetic properties to the recording process [5]. Also there exist sophisticated techniques and electronic circuits to detect and process signals. These factors make the magnetic recording the main technology of information storage at the present [3].

The main components of the magnetic recording systems are: First, the magnetic medium, which is usually a system of single domain magnetic (SD) particles, suitably prepared either as a tape or a hard disk, or metallic magnetic thin films [8]. Second, the write/read head, which usually consists of a soft core of ring shape or yoke, with a conductive wire wound around the core (this is called an inductive head). Another, advanced read head is based on the magnetoresitance effect (MR), which is suitable for perpendicular magnetic recording. Third, a spindle motor for moving the tape or disk, and an actuator for positioning of the read head. Advanced technological electronics and sophistical programs are used to encode and decode the signal and maximize signal-to-noise ratio (SNR) [7].

In the following, the basic magnetic properties of magnetic recording materials are presented, namely the hysteresis loop, and the time dependence of the remanent magnetization, which is essential for the writing and reading processes. Then, the recording media, followed by the writing and reading heads, and the recording process are briefly discussed. Finally, the heat assisted magnetic recording is addressed.

2. MAGNETIC PROPERTIES OF MAGNETIC RECORDING MATERIALS

2.1 HYSTERESIS LOOP

A schematic hysteresis loop appropriate for magnetic recording medium is shown in Fig. 1. The requirements for magnetic recording materials are high remanent magnetization (M_r) to provide high flux for a large signal in the reading process, and high enough coercivity to resist loss of recorded information due to stray fields or self-demagnetizing fields in the recorded pattern. The remanent magnetization is determined by the *squareness* of the hysteresis loop ($S_q = M_r/M_s$, where M_s is the saturation magnetization).

A material with high saturation magnetization and high squareness is therefore favorable for magnetic recording storage media. An ideal material with $S_q = 1$ should consist of a single magnetic domain crystal, which is not achievable on the macroscopic scale due to domain wall formation as discussed in chapter 1 of this book. One is therefore opt to resort to using a particulate material consisting of uniform single domain particles which can be aligned in a magnetic field. In any areal system, however, there is bound to be a distribution of particle sizes and anisotropies. So the size distribution should be as narrow as possible, with a mean size not too small, and not too large, a point which will be elaborate on in the context of discussing the thermal stability [10]. In fact, a value of S_q between $0.75 - 0.85$ have been achieved and predicted for isotropic material with random anisotropy [11]. Squarness means that the magnetization reversal occurs at a definite magnetic field value (Sharp transitions). The value of M_r, which is the magnitud of the magnetization at zero field, is the memory of that region of the medium to the magnitude and direction of the last field affecting that region.

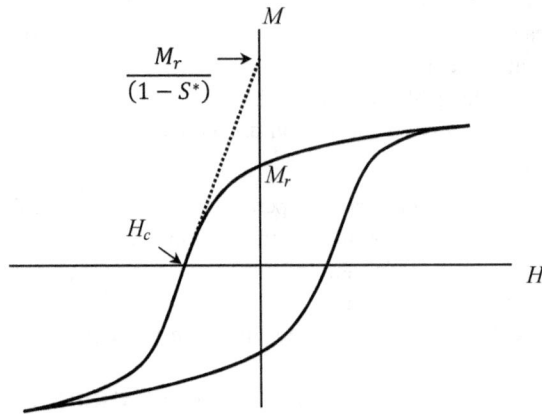

Fig. 1: Hysteresis loop of a ferromagnetic materials.

The coercivity (H_c) is also a very important parameter for magnetic recording, which should be moderately high to keep the recorded pattern against demagnetizing fields, but not too high in order to allow easier recording using the limited writing field available in a recording system [12]. The performance of a material as a magnetic recording medium, however, is lowered with the decrease of the coercivity due to thermal instability, which results in decaying of the recorded signal [13].

Another important parameter for characterizing a magnetic recording material is the *switching field distribution* (SFD), which should be narrow for high performance. The parameter (1 - S^*), which characterizes the width of the SDF, is the best and most accessible figure of merit for characterizing the recording material [10, 14]. This parameter is a measure of the slope of the hysteresis loop at the coercive field (slope of the dotted line in Fig. 1), which is defined by:

$$\left(\frac{dM}{dH}\right)_{H=H_c} = \frac{M_r}{H_c(1-S^*)} \tag{1}$$

A small (1 - S^*) parameter, and thus a high squareness, means sharp magnetization reversal at the coercive field, and a narrow SFD. The derivative of the $M(H)$ curve in the second and third quadrants of the hysteresis loop is shown in Fig. 2. It is obvious from this figure that the higher the peak value at $H = H_c$, the sharper the SFD, and the better the performance of the material for magnetic recording applications.

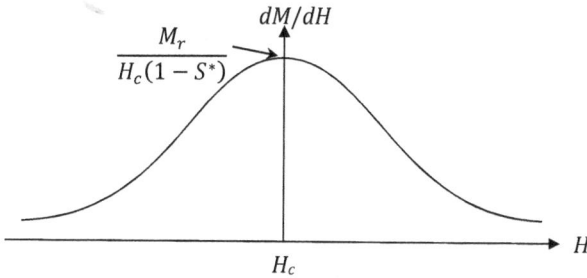

Fig. 2. Schematic diagram of the derivative of the magnetization with respect to applied field in the second and third quadrants of the hysteresis loop.

2.2   THERMAL STABILITY

The relaxation time (τ) of the magnetization characterizing SD particles (magnetization reversal) is given by the Nèel-Arrhenius relation [10, 12, 15, 16],

$$\tau^{-1} = f_0 e^{-E_b/k_B T} \tag{2}$$

Here f_0 is an attempt frequency, usually in the range 10^9 Hz – 10^{11} Hz, which was verified experimentally [17], and E_b is the height of the energy barrier. For a uniaxial anisotropy particle in the presence of an applied field H, E_b is given by,

$$E_b = KV \left(1 - \frac{H}{H_K}\right)^2 \tag{3}$$

where K is the anisotropy constant, V is the particle volume, and $H_K = 2K/M_s$ is the anisotropy field. It is relevant to mention that the application of H decreases the height of the energy barrier in accordance with Eq. (3). Eq. (2) is usually used to estimate the time required to store the data, and also to estimate the value of what is called the superparamagnetic limit and the *blocking temperature*. For simplicity, we let $H = 0$, then $E_b = KV$. For a given particle volume, if the relaxation time τ at a given temperature is greater than the experimental measuring time τ_m, then the particle is said to be *blocked*, where the thermal agitation is not capable of rotating the magnetization of the particle against the energy barrier. As the temperature increases, the thermal energy increases, and the relaxation time decreases in accordance with Eq. (2), until the relaxation time approaches the measuring time. The temperature at which the relaxation time becomes equal to the measuring time defines the blocking temperature (T_B) of the particle, above which the particle becomes *superparamagnetic*, where its magnetization vector rotates freely from one easy direction to another. Then according to Eq. (2), the blocking temperature T_B is given by,

$$T_B = \frac{KV}{k_B \ln\left(\frac{\tau_m}{\tau_0}\right)} \tag{4}$$

Notice that the exponential growth of τ with the parameter $x = KV/k_B T$ reflects the sensitivity of the relaxation time to the value of this parameter, and a value of $x \geq 50$ is required for storing the data for over a century in a real storage medium.

2.3 THE MAGNETIC AFTER EFFECT

In view of the finiteness of the magnetic relaxation time, the remanent magnetization after switching the field off is time dependent in accordance with the relation [18]:

$$M_r(t) = M_0 e^{-t/\tau} \tag{5}$$

Then for a sample composed of an assembly of uniform SD particles, substituting for τ from Eq. (2) yields,

$$M_r(t) = M_0 exp\left(-t f_0 e^{-KV/k_B T}\right) \tag{6}$$

As we mention earlier, however, the unavoidable distribution of particle size and anisotropy in a real sample results in a distribution of relaxation times $f(\tau)$. Accordingly, the remanent magnetization is given by:

$$M_r = \int M_0 f(\tau) e^{-t/\tau} d\tau \tag{7}$$

Notice that the exponential of the exponential behavior of the remanent magnetization given by Eq. (6) means that only particles with a limited range of particle size would have a significant contribution to M_r, namely, those with larger volumes. The time dependence of the remanent magnetization can be represented by the empirical formula [10, 12, 17, 19]:

$$M_r = C - S(H, T) \ln\left(\frac{t}{t_0}\right) \tag{8}$$

where C and t_0 are constants, and $S(H,T)$ is the magnetic viscosity coefficient. These relations demonstrate the decaying behavior of the remanent magnetization with time, which entails weakening of the recorded signal, and eventually loss of written information.

It is well established that the particle size in the recording material plays an important role in determining the performance of that material for magnetic recording. In view of the above discussion, the particles must be large enough to provide adequate stability of the recorded pattern for long times. On the other hand, requirements of efficient magnetic recording demand high signal-to-noise ratios, narrow transition regions between recorded bits, and smooth surfaces of the recording medium, all of which requiring a small particle size. Accordingly, the particle size should be optimized for high-density recording by a suitable compromise between these conflicting demands. In addition, time-dependent magnetic effects are manifested by the variation of the coercivity with time, which has an effect on the long-term stability of recorded information. The time dependence of the coercivity is given by the Sharrock's equation [20]:

$$H_c(t) = H_K \left(1 - \sqrt{\frac{k_B T}{KV} \ln\left(\frac{f_0 t}{0.693}\right)}\right) \tag{9}$$

At this point, it is relevant to mention that investigating the time dependence of the remanent magnetization involves saturating the sample, then switching field of, and subsequently measuring the magnetization as a function of time. These measurements contain different information from the remanent magnetization obtained from the M-H hysteresis loop as manifested by the following discussion.

The coefficient of magnetic viscosity (S), can be obtained by differentiating Eq. (8), where we have:

$$S = -\frac{\partial M_r}{\partial \ln(t)} \tag{10}$$

S was found to follow a distribution with maximum value near the coercivity [21]. Moreover, the time dependence of the magnetic behavior was described by the empirical relation:

$$S = \chi_{irr}\left(\frac{k_B T}{M_s V}\right) = \chi_{irr} H_f \tag{11}$$

where H_f is the fluctuation field, and χ_{irr} is the irreversible susceptibility. The fluctuation field was evaluated analytically for a system of aligned SD particles with uniaxial anisotropy, and found to be [22]:

$$H_f = \frac{H_K}{50}\left(1 - \frac{H}{H_K}\right) \tag{12}$$

Although this formula was derived for a fully aligned particle assembly, it remains of some value to the field of magnetic recording, since it was later shown that it gives the right behavior of S with magnetic field for a system of partially aligned particles, even though the magnitude of S was not correctly predicted by the formula [23]. The irreversible susceptibility can be obtained by differentiating (with respect to H) either of the *isothermal remanent magnetization* (IRM), or the *DC demagnetization* (DCD) curve. This parameter represents the SFD of the material [10].

The temperature dependence of the reduced remanent magnetization $\overline{M}_r = (M_r/M_s)$ reveals a distribution of blocking temperatures. The distribution function is derived from the derivative of the reduced remanence with respect to T [17]. Fig. 3 shows a set of experimental reduced remanence data for three different Fe_3O_4 fine particle powders, and the corresponding distribution functions (reproduced from reference [17]). Eq. (4) indicates that such a distribution represents the distribution of particle size of the magnetic particles as well as the distribution of their anisotropies.

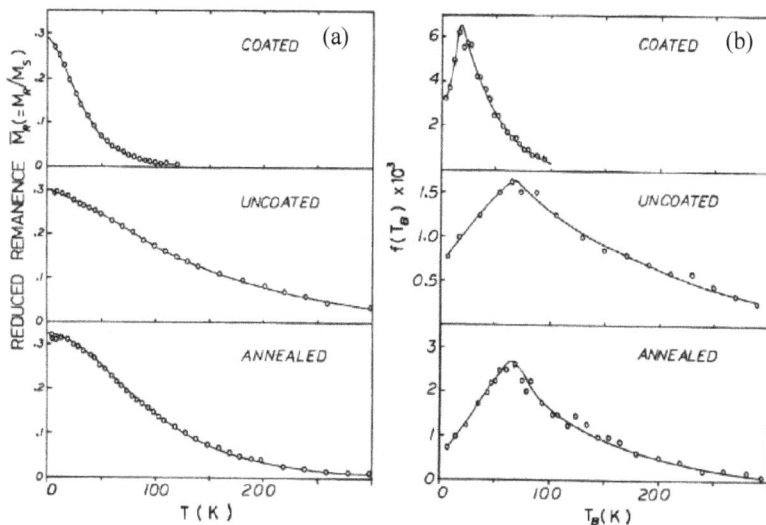

Fig. 3: (a) Temperature dependence of the reduced remanence measured at 100 s after switching the saturation field off, and (b) the normalized distribution of blocking temperatures ($\int f(T_B)dT_B = 1$).

3.   RECORDING MEDIA

There exist many kinds of magnetic recording media, some of which are simple and low cost, and some used for long-term information storage and backups. These types include hard disks, which are the main data storage media in computers, floppy disks and tapes, credit and identification cards, etc. Hexaferrites with high coercivity (M-type) are certainly useful for magnetic recording applications in which high coercivity is needed to inhibit erasure of recorded information without the need to rewrite on the medium, such as in the case of master tapes, and magnetic stripes on credit cards and badges. The high coercivity, however, poses technical problems in using the material for typical read/write applications. However, the capability of reducing the coercivity of the material by at least an order of magnitude as per discussions in chapter 3 and chapter 4 of this book, these materials remain of potential practical applications especially in the field of high density longitudinal and perpendicular magnetic recording.

Conventional magnetic recording on a disk or a tape is essentially the same. In both cases, a flexible or rigid supporting substrate is coated with a magnetic layer, which is set into relative motion with respect to the read or write head in the recording process. An additional upper layer for protecting the magnetic layer from wear due to contact with the read/write head, and to provide a smooth surface is usually added.

Perpendicular magnetic recording is an evolving application which provides the opportunity for increasing the recording density significantly. Hard disks for those purposes are usually made of different layers as demonstrated by Fig. (4). The first layer from the bottom is the substrate, which should be mechanically strong and stable, usually made of glass with a thickness of about 1mm. The second layer of thickness 1000 – 2000 Å consists of a soft magnet with high saturation magnetization, high permeability, and low coercivity. This layer is an additional layer on the conventional longitudinal recording disk, which is essential for guiding the magnetic flux from the recording pole to the collector pole. The third layer is the recording layer, which is the main layer in the disk; its thickness is about (200 - 500) Å. The magnetic particles in this layer should be isolated, to minimize the magnetostatic interactions and to improve the signal to noise ratio. The upper surface of the recording layer is coated with a protective layer (100 – 200) Å, and finally the lubricant layer (30) Å.

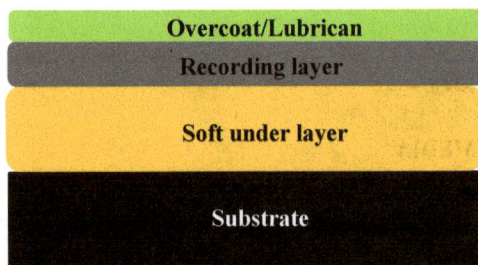

Fig. 4. Schematic diagram showing different layers of recording disk (not to scale).

The purpose of lubricant layer is to reduce friction between the surfaces and to reduce wear and other damages like erosion from air stream. Also, this layer protects the magnetic layer from progressive loss of material which may be caused by contact with the read/write head moving over the surface at very high speed. This layer is usually made of materials like Al_2O_3 or SiO_2.

To conclude this section, it is worth mentioning that in addition to the suitability of the intrinsic properties of the material for magnetic recording, the material should be stable under existing operating conditions, and the recorded information should be stable over long periods of time defined by the requirements of the recording application. Accordingly, several technical details of the design, as well as all relevant extrinsic parameters are equally important for the performance of a magnetic recording material [5, 24].

4. WRITING AND READING HEADS

Heads – contain all technologies concerned with the processes of writing or reading information in the recording media. The inductive writing and reading heads are similar (see Fig. 5); in fact, the same head can be used for both writing and reading in many devices. The head components are a soft ferromagnetic core in the form of a ring or toroid with a small gap, and a number of turns of conductive wire wounded around the ring (i.e. coil). A source of changing current (writing current) is used to feed the write head with signal representing the information to be recorded. On the other hand, when the reading head senses the magnetic flux of a magnetized region, an electromotive force is induced in the coil, and the variable signal detected by the reading head, therefore, represents the recorded pattern. This is simply the basis of the read and write processes in magnetic recording.

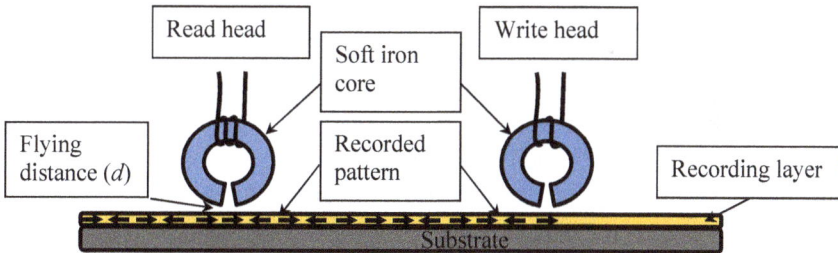

Fig . 5. Schematic diagram illustrating the principles of magnetic recording.

The output voltage in the reading head is proportional to the rate of change of the magnetic flux in the coil, in accordance with Faraday's law,

$$V = -N\frac{d\phi}{dt} \tag{13}$$

The signal amplitude depends on the relative speed between the media (tape or disk) and the head, and therefore this kind of heads (the inductive head) has limited sensitivity.

One of the requirements for increasing the recording density is the reduction of the transition region width (w) between magnetized regions. A relation for the transition region in a material with a square hysteresis loop was derived in terms of the recording layer thickness (t), remanent magnetization (M_r), head – medium distance (flying distance) (d) and the coercivity (H_c), which is given by [15],

$$w = \sqrt{\frac{M_r t d}{\pi H_c}} \tag{14}$$

To reduce w, either H_c should be increased, or the flying distance (d) should be decreased. The technical limitation to how small d can be imposes a limit to how we can use this parameter to improve the recording performance. Another restriction which limits reducing w is the difficulty of optimizing the ratio of (M_r/H_c).

Another kinds of reading heads is the magneto-resistive head (MR-head) [13, 25, 26]. The operation of the reading head material is based on the effect that the resistivity of certain magnetic materials (especially thin films), depends on the direction of the applied magnetic field relative to the electric current density (J). The resistivity of permalloy was found to vary according to the following equation:

$$\rho = \rho_0 + \Delta\rho\cos^2(\theta) \tag{15}$$

where θ is the angle between the magnetization in the head and the direction of J, ρ_0 is the resistivity of the isotropic material, and $\Delta\rho = \rho_\parallel - \rho_\perp$, is the difference between the parallel and perpendicular resistivity.

The magneto-resistive head detects the flowing flux through the head core, and has a higher sensitivity than the inductive head, since the effect is independent of the velocity of the tape relative to the head. Also the magneto-resistive head saturates in a small field (5-10) Oe [12, 15].

5.   WRITING/READING PROCESSES

During the writing process in the longitudinal mode, as the recording medium (tape or disk) moves past the head, a pulsed current flows in the wire wound around the soft core, generating a magnetic field. The fringing field from one edge of the gap penetrate the magnetic layer and closes at the other edge of the gap (it is easier for the magnetic flux lines to close through the magnetic layer than through the air). If the intensity of the fringing field is high enough to switch the moments of the magnetic particles, then the data are encoded as (One's). If there is no switching, the data are encoded as (zeroes). These transitions are synchronized with a clock, which synchronizes the magnetic recording motion. In the reading process, as the tape moves, the read head detects the induced magnetic flux through the induced voltage, which represents the output signal, i.e., the stored information.

Up to now, there are two modes of magnetic recordings. These are the longitudinal magnetic recording (LMR), in which the media magnetic anisotropy is oriented horizontally in the plane of the magnetic layer. The other mode is the perpendicular magnetic recording (PMR), in which the media magnetic anisotropy is aligned perpendicular to the plane [27]. The basic principle of operation of the write head in PMR is essentially the same as for longitudinal recording, but with a single pole writing process, where the magnetic field emanates from the pole perpendicular to the surface of the recording layer (Fig. 6). The magnetic flux lines penetrate recording layer to the soft magnetic under layer, and end at the collector end. The collector end has a much larger area in order to reduce the field strength in that end to levels insufficient to disturb the recorded pattern. The read head in PMR is usually a magneto-resistive head.

There are several advantages of PMR over LMR. Higher magnetizing fields can be used in PMR, allowing the use of a magnetically harder material, and greater stability of stored information. Also, the use of material with higher coercivity enables reducing the transition region width and the bit dimension, allowing for higher density recording.

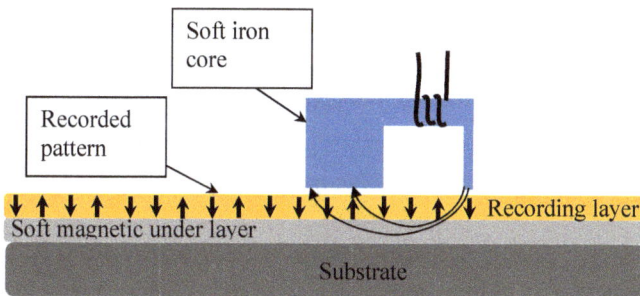

Fig . 6. Write head for perpendicular recording

6.   HEAT ASSISTED MAGNETIC RECORDING (HAMR)

There are some proposed techniques, like using electric fields, to overcome the thermal stability problem and to utilize materials of high anisotropy [13, 28]. But for now the main technique is the heat assisted magnetic recording (HAMR). This technique can be used for both longitudinal and perpendicular recording.

HAMR is a technique that allows the recording on high–anisotropy media, with a low write head field, by raising the temperature of the recording region. The temperature of the recording region is raised by a highly focused laser beam during the recording process, then rapidly lowered to room temperature to store the recorded information. The heating and cooling rates should be the same and comparable with the recording time ~ 1ns. This technique makes use of the reduction of the coercivity with temperature (Fig. 7). This allows the utilization of materials with higher coercivity for the recording media, which insures better stability against loss of stored information, and provides the opportunity for reducing the transition regions and therefore increasing the recording density. In addition, the reduction of the coercivity of the recorded regions by means of heating allows using a weak write field, which is usually one of the major challenges in magnetic recording.

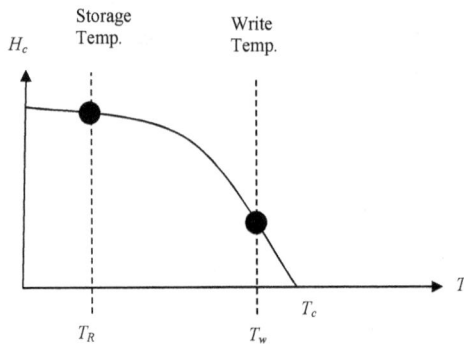

Fig. 7. Schematic diagram showing the variation of the coercivity with temperature. The Curie temperature (T_c), writing temperature (T_w), and reading temperature (T_R) are indicated on the curve).

REFERENCES

[1] S.X. Wang, A.M. Taratorin, Magnetic Information Storage Technology: A Volume in the Electromagnetism Series, Academic press, New York, 1999.
http://dx.doi.org/10.1088/0022-3727/35/19/201

[2] A. Moser, K. Takano, D.T. Margulies, M. Albrecht, Y. Sonobe, Y. Ikeda, S. Sun, E.E. Fullerton, Magnetic recording: advancing into the future, Journal of Physics D: Applied Physics, 35 (2002) R157-R167.

[3] N.A. Spaldin, Magnetic materials: fundamentals and applications, Cambridge University Press, Cambridge, 2010.
http://dx.doi.org/10.1017/CBO9780511781599

[4] The physics of ultra-high-density magnetic recording, Springer Science & Business Media, New York, 2012.

[5] G. Bate, Recording materials, in: P. E, Wohlfarth (Ed.) Ferromagnetic materials, North-Holland Publishing Company, New York, 1980, pp. 381-508.

[6] K.J. Strnat, Modern permanent magnets for applications in electro-technology, Proceedings of the IEEE, 78 (1990) 923-946.
http://dx.doi.org/10.1109/5.56908

[7] H.N. Bertram, Theory of magnetic recording, Cambridge University Press1994.
http://dx.doi.org/10.1017/CBO9780511623066

[8] J.C. Mallinson, The Foundations of Magnetic Recording, 2nd ed., Academic Press, San Diego, CA, 1993.

[9] M. Camras, Magnetic recording handbook, Van Nostrand Reinhold Co., New York, 1987.

[10] R. Chantrell, K. O'Grady, Magnetic characterization of recording media, Journal of Physics D: Applied Physics, 25 (1992) 1-23.
http://dx.doi.org/10.1088/0022-3727/25/1/001

[11] D. Jiles, Introduction to Magnetism and Magnetic Materials, Springer, New York, 1991.
http://dx.doi.org/10.1007/978-1-4615-3868-4

[12] R. Skomski, J.M.D. Coey, Permanent Magnetism, Taylor & Francis, New York, 1999.

[13] M.H. Kryder, E.C. Gage, T.W. McDaniel, W.A. Challener, R.E. Rottmayer, G. Ju, Y.-T. Hsia, M.F. Erden, Heat assisted magnetic recording, Proceedings of the IEEE, 96 (2008) 1810-1835.

179

http://dx.doi.org/10.1109/JPROC.2008.2004315

[14] E. Koster, Recommendation of a simple and universally applicable method for measuring the switching field distribution of magneting recording media, IEEE Transactions on Magnetics, 20 (1984) 81-83.
http://dx.doi.org/10.1109/TMAG.1984.1063006

[15] K.H.J. Buschow, F.R. Boer, Physics of magnetism and magnetic materials, Springer2003.
http://dx.doi.org/10.1007/b100503

[16] B.D. Cullity, C.D. Graham, Introduction to magnetic materials, 2nd ed., John Wiley & Sons, Hoboken, NJ, 2011.

[17] S.H. Mahmood, I. Abu-Aljarayesh, On the static and time-dependent magnetic properties of Fe_3O_4 fine particles: effect of oxidation, Journal of Magnetism and Magnetic Materials, 118 (1993) 193-199.
http://dx.doi.org/10.1016/0304-8853(93)90177-4

[18] S. Charles, J. Popplewell, Ferromagnetic liquids, in: E.P. Wohlfarth (Ed.) Ferromagnetic Materials, North Holland Publishing Company, Amsterdam, 1980, pp. 509-559.

[19] I. Abu-Aljarayesh, A. Al-Bayrakdar, S.H. Mahmood, The effect of heating on the magnetic properties of Fe_3O_4 fine particles, Journal of Magnetism and Magnetic Materials, 123 (1993) 267-272.
http://dx.doi.org/10.1016/0304-8853(93)90452-8

[20] M. Sharrock, Time-dependent magnetic phenomena and particle-size effects in recording media, IEEE Transactions on magnetics, 26 (1990) 193-197.
http://dx.doi.org/10.1109/20.50532

[21] K. O'Grady, R. Chantrell, J. Popplewell, S. Charles, Time dependent magnetization of a system of fine cobalt particles, IEEE Transactions on Magnetics, 17 (1981) 2943-2945.
http://dx.doi.org/10.1109/TMAG.1981.1061621

[22] R. Chantrell, M. Fearon, E. Wohlfarth, The time-dependent magnetic behaviour of fine particle systems, physica status solidi (a), 97 (1986) 213-221.

[23] S. Uren, M. Walker, K. O'Grady, R. Chantrell, Magnetic viscosity and switching field distributions in recording media, IEEE Transactions on Magnetics, 24 (1988) 1808-1810.
http://dx.doi.org/10.1109/20.11609

[24] T.C. Arnoldussen, E.-M. Rossi, Materials for magnetic recording, Annual Review of Materials Science, 15 (1985) 379-409.
      http://dx.doi.org/10.1146/annurev.ms.15.080185.002115

[25] S. Khizroev, D. Litvinov, Perpendicular magnetic recording: Writing process, Journal of Applied Physics, 95 (2004) 4521-4537.
      http://dx.doi.org/10.1063/1.1695092

[26] J. Heidmann, A. Taratorin, Magnetic Recording Heads, in: K.H.J. Buschow (Ed.) Handbook of Magnetic Materials, Elsever, Amstrdam, 2010, pp. 1-106.

[27] T.W. McDaniel, Ultimate limits to thermally assisted magnetic recording, Journal of Physics: Condensed Matter, 17 (2005) R315.
      http://dx.doi.org/10.1088/0953-8984/17/7/R01

[28] D. Weller, A. Moser, Thermal effect limits in ultrahigh-density magnetic recording, IEEE Transactions on magnetics, 35 (1999) 4423-4439.
      http://dx.doi.org/10.1109/20.809134

Keywords

About the Authors

Professor Sami Mahmood

Currently, Professor Sami Mahmood is professor of Physics at the University of Jordan, Amman. He obtained his B.Sc. in Physics from The University of Jordan, Amman in 1978, and Ph.D in Physics from Michigan State University, East Lansing, Michigan in 1986. He assumed several administrative positions at different Jordanian Universities such as Director of the Center for Theoretical and Applied Physical Sciences (1992-1994), Chairman of Physics (1996-1998), Dean of Science (1998-2004), Dean of Scientific research and Graduate Studies (2007-2009), Director of the Center of Accreditation and Quality Assurance, and Vice President (2004-2005). He has obtained several national, regional and international Awards and Honors for Academic excellence and contribution to science. To date he has published over 100 articles in peer reviewed journals and conference proceedings. Other academic and scientific activities included membership in the editorial boards of several national and international research journals, participation and management of nationally and internationally funded projects concerned with the establishment of new academic programs, capacity building, and program development, in addition to his active participation in scientific committees and councils of Scientific Research Funds in Jordan, and organization of national and international conferences.

Professor Ibrahim Abu Aljarayesh

Professor Ibrahim Abu Aljarayesh joined Yarmouk University in 1987 and became Professor of solid state physics in 1997-current. He optained his B.Sc. in Physics at the University of Jordan, Amman, in 1977 and his M.Sc. in Solid State Physics at the University of Jordan in 1980. He received his Ph.D. in Solid State Physics from University of Illinois in Chicago in 1986. During 2006-2012 he was Editor-in-Chief for the Jordan Journal of Physics. His research interest are: dynamics as well as statics of phase transitions especially in magnetic materials. He has published over 50 research articles in the field of magnetism and related materials.